MW00934235

Aircraft Battle Damage

Mike,

Thanks for your You Tube channel.
If you would review this book on your
channel, I would appreciate it. I know
you will enjoy the book.

Ian Freedom

For Aircraft Battle Damage Repair technicians and engineers
In the United States Air Force - a Dying breed.

To secure peace is to prepare for war
- Carl von Clausewitz

Disclaimer

This book is purely fiction. Any reference to a character
that is like a real life person is purely coincidental. No
classified information was divulged in this book. Any
actions by state actors are purely fictional and probably
did not happen.

CHARACTERS

Family and Friends
Betty - Steve's older sister
Linda Craken - Steve's Mom
Thomas Craken - Steve's Dad
Steve Craken - an above average dude that became an officer in the Air Force
Best Friend Kyle - Steve's best friend from middle school
Mat Hattier - High school friend
Phil - Dive buddy from college

Mississippi Air National Guard 72 MXS
SrA/TSgt Jordan Smith - Jet engine Mechanic
SrA Steve Craken - Jet engine Mechanic
SrA Brandon Lance - Jet engine Mechanic
SMSgt Vern Giles NCOIC of Jet engine shop
TSgt McCormick Test Cell manager
SSgt Rick Gernes Supervisor Jet engines

MSU Aerospace Engineering Class 1999
Gregg Barns - Student
Shane Opolis - Student
Christine Bates - Student

USS Memphis
Captain Jeffery Captain of the boat
Seaman Halls SONAR Operator

USS Tucson
Captain Miles Captain of the boat
Seaman Marianelli SONAR Operator
Seamen Trainor Fire Control Operator
Seaman Shand Defensive countermeasures

DDG-178 Japanese Destroyer JS Ashigara
Captain Yamamotto
Seaman Isk Fire control specialist
Crew of P3 Orion (Call Sign Black Sheep 3)
Lt Commander Smithey
Ensign Rands Plane tactical coordinator
Petty Officer Blake
Petty Officer Karls

E-6B Mercury TACAMO (Call Sign White Sheep 4)
Plane Captain Commander Rodriguez

People's Republic of China
Mr. Po Political Officer tasked with care, feeding and
spying on the ABDR team repairing the P3 on Hainan Island.
Maj Huin Sub Base Commander
Captain Wong recently appointed sub base commander
Ling Xu Captain Wong's wife

Kuái 1 Chinese next generation submarine
Captain (Shang Xiao) Wing Ping
Seaman (Shang Ding Bing) Xi Fire control operator

Kuái 2 Chinese next generation submarine
Captain Zhang
Seaman (Shang Ding Bing) Lang Sonar operator
Seaman (Shang Ding Bing) Phet Li Fire control operator

Chinese mini Sub attached to Kuái 2
Team Tiger
Petty Officer Fist Class (Shang Shi) Ping Commando Team
lead
Seaman (Shang Ding Bing) Li

Team Dragon
Seaman (Shang Ding Bing) Wong
Seaman (Shang Ding Bing) Pong Don

C/KC-135 System Program Office (SPO) Engineers
Col. Baxter System Program Director/Commander
Edgar Reels Head of C/KC-135 Structures shop and Steve
Craken's boss
Lt Jacob McFarland
Lt Rob Smelt (Josh)
Chase (Josh)
Joe Steve's coworker that sat next to him in the cube farm

Air Force Reserve Unit at Tinker AFB 507 MXS
Sgt McCay Maintenance Supervisor Aerial Refueling Wing

Tinker AFB

Major Baxter - Chief, Finance shop
Agnes - Gas station attendant, south side
SrA Bart Masters EOD technician

Oklahoma City Friends
Adam - A friend Steve met at church
Jewel - Steve's stalker girlfriend
Jerry - A Friend Steve met at church
Ronda - A friend Steve met at church
Jenna - Steve's godly girlfriend
Mac - Steve's church friend

ABDR Engineers
Capt Evans ABDR Engineer's Course Instructor
Lt Steve Craken Call sign Engineer One
Lt Tiffany Jones
Lt Geronimo (Gerry) Ford
Lt Mike Carry
Lt Sarah Timmons
Lt Adam Smith
Lt Lisa Swage
Lt Julie Bear
Lt Carly Roach
Lt Jacob McFarland

ABDR Headquarters Wright Patterson AFB, OH
Hazel Macintyre Point for all ABDR Engineers

654 Combat Logistics Service Squadron Tinker AFB
Lt Colonel Jones Commander CLSS
SSgt Baletti Training NCO
MSgt Bans C/KC-135 Team Lead
TSgt Bosk C/KC-135 Structures
TSgt Smith Electrician
TSgt Kaufman Structures
SSgt McMasters Hydraulics
TSgt Susan McCormick B-1 Assessor/Electrical
MSgt Gleeson B-1B Team Lead
MSgt Malcolm NCOIC of Training ABDR Training site
TSgt Simmons B-52 team lead

Diego Garcia (Deployed Location)
Col Bolls - Base Commander
Major Huffines - Coffee steward for Col Bolls
Petty Officer Chavez - Tower supervisor

Seaman Beaumont - RADAR operator
Amn Johnson National guard exchange member - Jet engine
mechanic
TSgt Helms - Civil Engineering Specialist
SSgt Combs - Aircraft Fire fighting specialist
Airman Sikes - Female Security Forces specialist

MacDill AFB FL
TSgt Sam Hurls - Structures troop
SMSgt Ford - Structures shop supervisor

Guam Air Base
Air Force General Deacon - Acting Pacific Commander
Major Splint - Patriot battery commander

B2 Crew
Spirit of Mississippi
Capt Ricks - Pilot
Maj Zillinger - Aircraft Commander

Spirit of Kansas
Capt Ware - Pilot
Capt Valance - Aircraft Commander

Schriever AFB
50th Space Wing
Col Susan Vaughn - Base Commander

23 Space Operations Squadron (23 SOPS)
Capt Hanes - Shift Commander
Airman Jones - Battle Damage Assessment Operator
Airman Rodriguez - Battle Damage Assessment Operator
MSgt Johns - NCOIC ops floor

Quanzhou China, Chinese Army/Air Command and Control
Ying Bi - Commander of Air Defenses Quanzhou area defense
Capt Ric - Fighter Flight Lead Sukhoi Su-30MKK

Okinawa Kadena Air Base
Capt Sinclair - Maintenance Officer

Uzbekistan
2nd Lt Lisa Cranston - Public Affairs Lt
Captain Price - Maintenance officer on C/KC-135s
Captain Ricks - Medical Doctor...Med Group

Col Smith - Base Commander
Maj Jim Forstal - Security Forces Commander
MSgt Powell - Structures Shop Supervisor
SRA Tidwell - Structures troop
A1C Walker - Structures troop
Lt Wilson - F-15E Pilot

Part One:

Experiences

1.

Friday June 2, 1989, 0600 Hours
Beijing, China

     Xu Genghis woke early in the morning to go to his

state assigned job in a local factory making cheap plastic

crap generally destined for America. He lived with several

family members in a one room apartment on the east side of

Beijing. Genghis and his family shared one bathroom with

six other families and he had a grand total of four minutes

to do what he needed in the mornings in that one small

bathroom. He worked 10 hours a day with a thirty minute

lunch break seven days a week. To say he was

disenfranchised with his country and communism was an

understatement. He watched his father, Xu Hu, work his

whole life in one menial job after another and finally

become ill with lung cancer from the asbestos at the ship

yard. Hu couldn't work any longer so the government cast

him aside like garbage.

Hu had named his son Genghis which means "Just and

Righteous". The family name of Xu was a common name in

China. Genghis felt he would never live up to his name.

He was no hero. In his short time allotted to eat

breakfast, which consisted of a handful of sticky rice and

some form of potted meat, he talked with his sick father

about the dismal government. Hu talked of the protest the

university students started a month ago. The protest was

based on greater political freedoms. Hu mentioned he

thought it was gaining some serious attention. Genghis had

been following the protest very closely himself but did not

tell his family.

Genghis was about to leave to ride the family bike to

work when the flimsy door was kicked in by the Chinese

elite guards. Everyone was forced to the floor with SKS
rifles pointed at their heads. Hu asked the guards what
they wanted and they told him to, "SHUT UP OLD MAN, YOU ARE
A TRAITOR TO THE SATE". Hu wasn't but they didn't care
because they had a man with the surname Xu that they were
looking for that was giving information to the Americans
about China. Specifically state sponsored internet attacks
and hacks on DOD computer systems. Hu didn't know any state
secrets. He was bound, gagged and dragged off. The family
never heard or saw of him again. The elite guard was in
the apartment and gone within five minutes. Genghis' mother
was hysterical but told him to leave for work or they would
haul him off like his Dad tomorrow for being late.

Genghis left the apartment and started riding his
bike. The blood was boiling in his veins. *They just took
his ill father for no reason or explanation. They had to be
stopped but how*? He thought.

He usually rode north and then west to the factory but
today he saw smoke to the north and thought it may have
been a fire which could make him late. The events this
morning almost made him forget he had to go by the market
on the way to work or his family would not eat that night.

The market routinely ran out of food by three in the
afternoon.

Genghis rode to the market and picked out a pigeon, a
small bag of rice and two durian fruits. The market lady
knew him and also gave him groceries for two other families
that lived in the building he lived in. The market was
abuzz with talk of the protest and Genghis started asking
questions. When, where, he asked. The general consensus
was for everyone to meet at Tiananmen Square this morning!

Thursday June 2, 1989, 0645 Hours
Tiananmen Square Beijing, China

Genghis arrived to the square on his bike laden with
groceries and decided to join the protest. He leaned the
bike up against a fence and grabbed the groceries so they
wouldn't be stolen. Food was a scarce commodity in China.
He joined a large group of protestors close to the main
road.

The students were the boldest people he had ever seen
in his life. He joined in their chants and he became alive.
He had never felt this way in his life. After an hour of
being in the protest someone shouted the People's Republic

of China (PRC) army was coming with a column of tanks. You could smell the fear but few people left. When the column came closer people started to flee for their lives. One protest leader called for the students to block the road and to not allow the tanks to pass and break up the protest. No one made a move to block the tanks… except Genghis.

He ran to the lead tank on the road and stood fast in the breach of freedom. When the tank tried to go around him, he would move to match it. This went on for an hour and the tank crew didn't know what to do because they knew the world was watching and the cowards in the Politburo (decision making entity of the Chinese government) were also watching. The leaders in the Politburo thought the crowd would run at the sight of the tank column. They did not. Since Genghis had his grocery bags with him he became known to the world as… Walmart bag guy.

Figure 1 Ghengis in Defiance

Finally the politburo ordered all television recording stopped. When the cameras were shut off they took Walmart bag guy into custody, bound him and beat him with batons. His family's groceries and the other families' groceries, Five day's wages, were smashed by the advancing tank column.

Friday June 3, 1989, 0930 Hours
Beijing, China

PRC Infantry and armor poured into the heart of Beijing and opened fire on the protesting civilians. The West knew what they had done and confronted China. They tried to cover it up saying, "The protest turned violent and we had to maintain order". There is no accurate death toll for the protest. China claims a couple hundred were killed but other sources claim much higher number in the thousands. Others were arrested and never heard from again. They were held as political prisoners in their own country.

The Politburo inferred a deep shame from the world that they could not maintain order without killing their own citizens and they vowed to build their military might to become a dominating force in the Pacific theater and then the world. They would wait for the perfect time to strike the West… Especially the United States with their pious tradition of freedom.

2.

Friday May 24, 1991, 11:23 P.M.
Starkville, MS

Steve Craken was in his junior year of high school and
had learned that living in Starkville MS was BORING! He
moved there with his family from Phoenix Arizona after his
5th grade year. He loved AZ. He loved the outdoors
especially hiking in the desert and sledding in the

mountains of Flagstaff. In contrast Mississippi was damp
all the time and the allergies were ridiculous. He joined
the boy scouts but really there is no hiking in MS. Sure
you have national forests but it's all the same… pine tree
over there pine tree right here, pine tree next to the
house.

    Steve was an adventurer so he made up his own stinking
adventures. He taught himself repelling from an old 75
foot oak tree in his front yard, which he found out later…
he was highly allergic to. He would climb up the backside
and tie his rope up on a higher limb and then go to a lower
limb and swing off. His black Labrador, Bart, would try to
bite his butt while he was swinging.

    One day Steve had some friends over from church and he
was showing off on his tree. He decided to rappel going
face first out of the tree while everyone was watching.

    When he ordered his gear from a catalog (two months
ago) they had regular carabiners and locking carabiners.
The locking biners were four dollars more so he went with
the regular one. Little did he know that the rope could
come across and unlock the regular biners and that time in
the tree it unlocked! Steve was sent plummeting to the
ground with a death grip on the rope. He landed on his

chest and bounced twice. He received a nasty rope burn on

his bicep but no other real injuries except his pride. His

Mom, Linda was cool about the whole issue. She let Steve do

mostly what he wanted and hey there were natural

consequences. Although Linda never knew about the water

tower or MSU stadium that Steve would attack next.

After a while the tree became routine. So he started

thinking. There are no cliffs in MS but there are man-made

things that are taller than the oak tree. There are fire

towers, the forestry service used to use to locate forest

wild fires thirty years ago, there are water towers and the

biggest of all… The Mississippi State University football

Stadium.

The West tier of the Mississippi State Stadium – The

Home of the Bulldogs, was roughly 150 feet high with a huge

spiral ramp leading to the upper tiers. So one night

around 1 A.M. his best friend Kyle and a few other crazy

friends sneaked into the stadium, walked up to the upper

tier and tied an anchor to a huge concrete pylon. Since it

was dark they could not tell if the rope touched to ground.

So Steve walked down the ramp a little until he could see

the ground and noticed the rope was not touching the

ground. He went down further and accessed the rope was

about 15 feet off the ground. The rope they were using was
a certified climbing rope and not a static rope typically
used for rappelling.

The difference in a climbing rope and a static rope is
a climbing rope is made to stretch so if a climber falls
the rope stretches a certain percentage and the climber is
not hurt in a sudden arrest. In contrast a rappelling rope
is made to take more abuse. Basically the person is sliding
down the rope with a mechanical system. The person
rappelling does not want any stretch in the rope.

Steve decided that since he tied the anchor and the
rope didn't touch the ground he would go first. Steve put
on his harness, threaded the rope through the descender and
locked the descender to his harness with a locking
carabiner this time.

There was a slight breeze blowing but up over 100 feet
is was fairly intense blowing the rope around. Steve had
his buddies check out his gear and hopped over the rail.
He looked at everyone and said, "HERE GOES EVERYTHING",
then he kicked off the tier and started going down the
rope. There were two cross tiers and then a 75 foot free
fall. Steve hit the second tier with his feet in the "L"
position, used for rappelling and then the free fall. He

stopped and looked at the concrete ground 60 feet below him

and saw the rope was still not close to the ground.

There was no going back up (he hadn't bought ascenders and

that would have taken over an hour to climb back up) so he

took it slow descending down the rope. He finally put his

feet on the ground with 3 inches of rope on the slack side

of the descender. The rope had a tension of 140 pounds

(what Steve weighed) on it. Steve had some really good

common sense but sometimes he chose to ignore it. This was

one such occasion. He let the three inches of slack rope

go through the descender and the rope flew up like a rubber

band under tension. In the process it slapped the mess out

of Steve's neck. He didn't care because, hay, he just

rappelled off the MSU football stadium in the middle of the

night and didn't DIE or get caught!

Several weeks later Best friend Kyle and Steve decided

to go back to the MSU stadium to rappel just the two of

them. Kyle was wearing his Florida State Seminoles bright

yellow and red wind breaker and Steve said, "Are you

wearing that tonight?" Kyle replied, "Yea we never get

caught." Steve was wearing his typical all black attire.

Well they were both feeling cocky and maybe lazy so Steve

parked the Blue Ford directly where they would be

rappelling in a driveway at the stadium. They got out of the car, hopped the fence and started walking up the long ramp that would take them up to over 150 feet to the upper tier so they could begin tying the anchor. They both saw the car's headlights turn into the same driveway they just parked at. They went prone on the large pedestrian ramp.

A loud radio call was heard about half way up the ramp. "**7 David 3, License Plate Oktibbeha CA6335 Registered to Thomas Craken address 103 Hoover Drive Starkville MS**".

Steve looked at Kyle, Kyle looked at Steve. They mouthed the same word…"Crap". About that time the ramp lit up like daylight. There was so much light that Kyle's Florida State jacket was turning the rest of the ramp a bright orange. "We never get caught, huh?" snorted Steve. "Shut up," came the reply. Their only hope was to low crawl up the ramp to the first section and exit on the far side of the stadium but then the world returned to the dark. The police search light then shown across the field to the visitor's side. They stayed put for a few minutes and the light went out and the police car left. No words were said just two teenagers running as fast as they could down the ramp to get out of there before there were more police on the scene.

3.

Friday March 8, 1991, 8:45 P.M.
Starkville, MS

Cruising west on Highway twelve. Kyle looked at Steve

and asked, "Cat Explosion?" whereas Steve replied, "Oh yes

it is time. When did we do the last one?" "I don't know

three weeks ago." Came the reply from Kyle. The two drove

over to where Kyle lived at the Pine Manicured Home

community or the local trailer park. Steve turned onto

First Pine Street and cruised past the pool to Kyle's trailer and pulled over a little past it. Kyle jumped out and went to the back side of the trailer and banged three times on the wall. He then ran to the car and they left in a hurry. They drove to Steve's house because Kyle was spending the night.

4.

Friday March 8, 1991, 2:14 A.M.
Starkville, MS

The water tower experience gave them the confidence

for the MSU stadium. The water tower experience consisted

of two attempts. The second attempt was glorious. Here is

the account of the first: The time was 0214 on a random

Saturday morning. Steve and Kyle probably snuck out of

Steve's house. Steve got in the baby blue ford Escort and

gently closed the door. Kyle opened the passenger door and

began to push the car backwards down the drive way. The two

hundred year old plantation house sat on a slight hill with dug-out dirt roads.

In the cotton picking era of Mississippi they would build the dirt roads by digging out the slight hills so the cotton wagons didn't go over any hills on the way to the gin. If they went over any hills the cotton could fall off the wagon because they would pile the wagons with high cotton and they would lose money.

When the Escort built up some speed Kyle jumped in and closed the door. At the bottom of the drive Steve let out the clutch (popped the clutch) and the momentum turned the engine and the key was "on" so the engine fired. They drove to the water tower across town at the corner of Whitfield and Scales St.

Kyle and Steve chunked their gear over the fence and hopped over it. The fence had the strait barb wire at the top which is really a waste of money. It may look menacing but is not very hard to get over. They went to the tower leg that had the ladder on it. The ladder started at seven feet off the ground to prevent people like them from gaining access to the tower. Steve clasped his hands together and Kyle put a foot in his hands and Steve lifted Kyle up to the bottom rung of the ladder. When Kyle was on the ladder, Steve threw the gear up to Kyle. Then when Kyle

was twenty feet up the ladder, Steve ran and kicked off the

tower leg and jumped up to grab the bottom rung.

This tower had zero safety gear. Usually the ladder

would be surrounded by a metal skeleton tube so if the

climber fell back the tube would catch them or the ladder

would have a safety cable you could clip into it wearing

your harness. The cable would have catches every ten feet

so the most you would fall is 10 feet before being arrested

by the cable. Also this ladder had Mississippi engineering

at the top. The tower legs have a wider base at the bottom

and smaller circumference at the top. They are angled in

towards the circle that meets the tank at the top. So the

climber has a positive slope going up instead of purely

vertical. When the ladder comes up to the cat walk there is

supposed to be a hole cut into the cat walk and the climber

climbs up through the floor of the cat walk and its

relatively safe especially if the climber is clipped onto

the cable.

Well none of that safety stuff was on this tower so

when the climber got to the top, and was tired, he had to

climb ten feet of a NEGATIVE SLOPE, where they are actually

hanging backward and climbing up at the same time because

they didn't cut a hole in the cat walk they just attached the ladder to the outside rail and the climber had to climb up and over the rail. It was sporty.

Kyle and Steve would have to wait until next time to find out about the lack of safety equipment on the tower because Steve was trying to find some purchase on the slippery tower leg and accidentally kicked a four inch metal pipe that ran down from the tower to about six feet off the ground. Maybe it was an overflow pipe, who knows but it was only attached at the top of the tower and in effect it became the world's loudest gong! **GONG GONG GONG GONG** Dogs were barking, house lights were coming on, TIME TO GO, Steve said into his headset radio to Kyle. Kyle said, "Elevator one going dooooooooown." Every dog within two miles was barking at the gong.

5.

Saturday March 9, 1991, 11:15 A.M.
Starkville, MS

Steve dropped Kyle off at his trailer and went back

home at around 11:15 A.M.

Kyle walked in the front door and June, Kyle's Mom

said, "Well you missed it last night!" Kyle replied, "What

are you talking about?" June said, "I was almost asleep

watching Letterman on the couch. At midnight there was a

cat explosion!" A cat WHAT?" exclaimed Kyle trying to keep

a straight face. "Woodstock started running for his life

and he climbed the wall right there where you're standing!
Samantha ran straight to my room and still hasn't come out
from under the bed and Sasha jumped down on the floor,
froze up, turned around looked at me, meowed and then ran
to your room. Poor Woody and Sammie are still hiding. I
think it was an earth quake. You know those drills you have
at school for the Memphis fault?" "Man that is crazy."
Replied Kyle and went about his day.

6.

Friday August 16, 1991, 01:33 A.M.
Construction site at MSU, MS

Best friend Kyle looked at Steve and said, "This place is boring. Let's go do something." I saw a construction site on MSU and its pretty dark, replied Steve. "We're there!"

The construction site that would later become the MSU forestry building was a modern brick building and when they finished it a student noticed there wasn't a stick of wood on it so they had to go back and put in some wood beam

accents to the roof. - Mississippi engineering.

Steve parked his baby blue Ford about three blocks

away from the site. They watched for cars coming but the

town was dead that late at night. The site was far away

from the fraternity houses and the parties there. They

gained access and started exploring the massive building.

It was fun. The two teens imagined they were in a war torn

country executing a search and destroy mission. In one room

they found some engineering drawings of the building and

there were some sweet sticks of wood used as paper weights.

It was a 2 X 2 by three feet long. Steve grabbed one to

take with them and asked Kyle if he wanted one. Kyle said,

"No", Steve asked him no less than four more times because

they kept running across them in different rooms. Kyle

still said, "No".

They found an open stairwell that had yet to have the

bricks installed and they had their rappelling gear so they

went ahead and anchored up and did some SWAT style rappels

down the stairwell. On the second run Kyle spied a cherry

Picker.

A Cherry Picker is a four wheeled crane with a basket

on top to allow two operators to be lifted in the air to

where ever they needed to go to aid in construction. They quickly packed up the gear and went to investigate. They climbed in the lowered bucket and looked at each other and said, "The key is in it!" Kyle cranked it up and it started right away. It was loud but there was no one close to the construction site. Although the engine was running, it would not move. The drive mechanism would not move the wheels nor the crane controls to move the crane. They shut it down and cranked it up again when Steve spied the safety foot control! He jammed his foot down on the foot pedal and the picker came to life. Kyle had the crane controls and Steve had the cart drive controls. It was awesome until Kyle started yelling about some ditch. Steve was like we are going through the ditch but Kyle stopped him because they were twenty feet up and probably would have catapulted themselves through the air.

The next day Steve woke up late and didn't have anything to do so he called Kyle and asked if he wanted to go to Walmart. He said he did so Kyle came over to Steve's house. Steve had been whittling on the stick they got the previous night. It made a nice weapon and Kyle said, "That is cool I want one." Steve just looked at Kyle while shaking his head.

They left and went to Walmart. They were messing around the store when they came to the toy isle. There was some little boy playing with a cap gun. He was aiming it at Steve and Kyle. Steve picked up a ball and asked the kid if he wanted it. Steve was going to be rude when the kid said yes then Steve would pull it back and say something like "well find your own." The little kid pushed Steve's hand away and told him, "No, I'm playing cops and robbers."

Just then Kyle said to the little boy, "Ah you like playing cops and robbers?" The boy said, "Yes.", in a matter of fact tone. Kyle said back to him, "I used to like to play that but now I am all grown up and like to play Pimp and Hooker." Steve was like, **"DUDE!"** About that time the little kid's Mom rounded the corner hearing the new game that Kyle liked to play and that was exit stage right for Kyle and Steve.

7.

Monday July 12, 1993
Starkville, MS - Home of Mississippi State University - The
Bulldogs!

Steve was living large! He had zero cares in the
world. He was living at his parents' home and just started
his freshman year at Mississippi State University in
Aerospace engineering-Summer School. He had just completed
Air Force Basic Military Training and Technical School at

Chanute AFB Illinois on Jet Engines. As far as he was
concerned he knew everything and he was awesome.

Steve's family had lived in a big plantation home that
was over 200 years old on a secluded 10 acre lot. The house
was close to MSU situated behind the Sac N Save grocery
store. You could not see the house if you drove by the main
road because of all the pine trees. It sounds really
picturesque but there was no central heat or air
conditioning so it was hot in the summer and cold in the
winter. In fact the only heat for the living room and
kitchen was a wood burning fireplace.

Bart was the three year old black Labrador family dog.
He was cool. One night Steve came home from his job at Mr.
Cooks. Mr. Cooks was a sub-standard burger joint close to
the MSU campus. Steve worked the front register and took
out the garbage on occasion making the local raccoon
population scatter from the dumpster. When he pulled onto
Hoover Drive (where his house was) he saw that his family
didn't bother leaving the porch light on for him. He parked
his baby blue Ford and started walking toward the front
porch. It was close to midnight and very dark. The next
thing Steve knew was that he was on the ground. *What the
hec just happened*? Steve thought. Then he could hear his

lab panting and could smell him. Bart had tackled him. He was jet black so you could not see him. HE WAS WAITING TO TACKLE ME! The next night Steve was watching and listening for him and here he came to tackle him again but Steve was ready for him. On occasion Bart would come up with all manner of weird "food" items. Steve's family deduced that the dock workers at Sac N Save were giving him old food. He came up one day with a whole sea bass in his mouth about 14 inches long! Then one day he was chewing on some kind of joint that looked like a knee joint. The next day he had an old woman's shoe. Steve said to Linda, his Mom, "I think he got the little blue haired lady down the street."

That house was cool to live in. One day the front porch light shorted out and caused a fire because a squirrel inside the wall was chewing on the wire. It was just after Steve graduated from high school. Steve would go weeks without seeing his sister, Betty… so they had a great relationship.

A friend of the family rented her Dad's old house to the Craken family for a while. The family went from 3,500 square feet living area and 10 acres to 1,100 square feet and 1 acre on a busy road! There was one bathroom in the tiny house. Steve and Betty would go round and round about

sharing the bathroom, sharing the carport, sharing the
laundry. Basically Betty didn't share.

Betty had an old crotchety tabby cat named
Butterscotch and Steve had Bart the lab. You had to watch
your stuff because that old tomcat would get in your
laundry and pee in it. There is no way known to man to
remove the smell of cat pea. Steve had to burn those
clothes. Steve and his sister went at it and so did Steve
and that cat.

One day Steve came home from his classes and saw the
old cat sitting in the drive way so he gunned his jeep's
engine and chased the cat into the back yard. Apparently
Butterscotch did not like that because the next day Steve
got in the jeep and discovered the cat had peed up the
center console so Steve did his best to clean it up and
that made him late for class. That day when he came back
from class he really gave the old cat a run for his money
and the next day he would have a fresh cat pea smell in the
jeep. That pattern continued a couple of days. Then the cat
escalated the whole war by crapping right on the center
console lid leaving Steve a gift of epic proportions. Steve
was vexed to say the least.

Steve came back that day from class and he did not see the cat so he was a little disappointed he couldn't chase the cat and really by this point give it a heart attack.

He parked the jeep and went to let Bart off his run and play with him a little before hitting the books. When Steve let Bart off the run he went straight over to a bush, which was strange for the lab. Typically he would go around the whole yard sniffing things and playing. So Steve thought there may be some woodland animal that needed rescuing but when he got over to the bush Butterscotch was sitting under it. As soon as Steve approached the duo, Bart raised his leg and pissed all over that cat! Butterscotch was deeply offended and tried shaking it off his paw to no avail. Steve got right up in the cats face and exclaimed, **"That's what you get you filthy animal!!!!"** The cat never set foot inside Steve's Jeep again.

8.

Saturday September 10, 1993
Jackson, MS Thompson Field Air National Guard Base.

Senior Airman (SrA) Steve Craken was sent to the flight line with SrA Jordan Smith to help out the flight line technicians deal with a problem on a C-141 Starlifter's #2 jet engine during a normal schedule drill weekend. Both Airmen were jet engine mechanics in the shop and had a couple of years' experience repairing jet engines. When they got to the jet, one of the TSgt (a middle enlisted rank) mouthed off and said, "We don't need any Airmen messing things up". So SrA Craken said, "OK"

and just walked back by the truck. Then SrA Craken went

over to the TSgt's tool box, while he wasn't looking. It

was sitting on the ground. Steve grabbed it and rolled it

several times so all the tools would fall out of their

assigned places within foam inside the tool box. It would

take the TSgt maybe an hour to put all the tools back where

they belonged in the shadow foam at the end of the day.

9.

Monday 1800 August 15, 1994
Colvard Student Union

SCUBA Diving is awesome. Steve had always wanted to become a diver but he lived in Mississippi about five hours' drive to the coast and that was the Mississippi coast where the beach is mud and there is zero visibility in the murky brown water. Then Steve saw SCUBA lessons for $150 on a bulletin board and jumped at the opportunity.

Tonight's class, that Steve was late for, was on sea life and the potential dangers they possess. There was one guy…You know that guy that was asking the stupid questions. There was a professor that said, "There are no stupid questions…just stupid people. This guy was one of those. His name, the Name that his mother gave him was actually Bubba. It was not a nick name but actually Bubba Gilliam. Steve didn't know his real name so he labeled him "Mr. No Life Skills". Bubba asked how he could fight off a hammer head shark or how to catch a barracuda with his bare hands. *How did this guy make it to college?* Steve thought. The Instructor answered his questions fairly tactfully relying on stories from his experience. Steve enjoyed the class and determined that most sea life cannot hurt you if you are wearing gloves and a wet suit but still it would be cool to have a dive knife.

10.

Saturday August 20, 1994
Jackson, MS Thompson Field Air National Guard Base.

Mississippi is hot and humid in the summer. Drill started fairly regularly with SrA Craken going into the shop before roll call at 0720. The shop was empty. There were no engines on the maintenance stands being worked on. That means a long boring drill weekend. Steve thought, *Oh man I'll have to be studying the Jet engine advanced course for my Five Level. That stinks.*

The maintenance group had roll call in the big hanger on the south side of the base. Steve greeted all his friends and walked to the engine shop. Everyone was in the

break room and the supervisor came in and asked for

volunteers. Usually you don't volunteer for anything in the

military but it had to be better than his correspondence

course and maybe it was something cool like cleaning the

test cell. "I'm in", said SrA Craken. SMSgt Giles said, "I

need a team to go out and perform an engine test run on one

of the home birds", Steve said, "OOOOh that is cool".

"You'll need a speed handle with a 7/64 Allen wrench bit on

the end to trim the motors". Instructed the SMSgt. SSgt

Gernes told the SrA to be the trimmer. They drove out to

the flight line after getting their flight line badges and

got set up to run the number two engine. Aircraft engines

are numbered from left to right as looking from the rear of

the aircraft. SSgt Giles instructed Steve on how to

approach the engine and the dangers associated with

trimming engines while they are running with 33,000 pounds

of thrust out the back end and the possibility of sucking a

man into the intake in the front. "There are two safe ways

to approach the underside of the engines on a 141. You can

come from the wingtip and walk directly under the outboard

engine and then to the inboard engine or the second option

is to come from the nose of the aircraft and stay as close

to the fuselage as possible and then walk under the number two engine. You do have to watch out for the hot APU exhaust located in the left wheel well. When you are directly underneath the engine you also have to watch out for the starter/generator exhaust. The exhaust flows down and unfortunately it is directly next to the trim screws but it is not as hot as the APU exhaust. Basically you need to act like a scared cat in a room full of dogs." said the SSgt. Steve said, "OK scared cat…got it."

About thirty minutes later the crew started the engines. The two left engines needed to be trimmed. They were recently replaced by rebuilt engines. The right side were already trimmed months ago when they were installed on the wing.

SSgt Gernes put on a headset that was plugged into the com panel of the jet just inside the right crew entry door. Then he crouched along the fuselage by the APU and then outboard under the number two engine. Steve had already donned ear plugs during the truck ride out to the flight line but when he got out he grabbed a set of ear defenders and put those on as well because he knew SSgt Gernes was almost deaf and Steve wanted to keep his hearing.

Steve followed the same path as the SSgt. He quickly found out what the SSgt had meant in regards to the starter/generator exhaust. He was wearing his long BDU shirt but still felt the heat. He had to scrunch up his arms to get into position to be able to screw the trim screw. On the TF-33 engine there are two trim screws, a rough trim and a fine trim. They started with the rough trim screw and SSgt Gernes gave hand signals to turn it counter clockwise three turns. Then they waited a few seconds for the instruments to react in the cockpit. He then gave hand signals for one turn clockwise. That was all it needed. The fine tuning screw was left untouched. The same drill was accomplished on the outboard number one engine with similar adjustments. When they were finished Steve had a massive headache but he thought, *how cool was that being under 33,000 pounds of thrust trimming a motor?*

11.

Saturday 0900 September 3, 1994
Industrial Complex of Hwy 278 Oxford MS

When the SCUBA class started everyone was stoked. The students were dreaming of a cheap beach trip to Florida to get certified seeing all kinds of sea life and hanging out on the beach and eating some great sea food. They all assumed they would be doing the underwater training in MSU's Olympic size pool.

Steve had some friends when he was in high school that had parents that worked for MSU so they had memberships at the MSU pool for the summer. It was a cool place to hang out especially with hot college girls in bikinis! Unfortunately for Steve the cost for non-students/faculty was $100! He didn't have that, his parents didn't have that so he was forced to pay $5 every time he went. He couldn't afford that very often either. Then a family started coming to his Dad's church and they had college kids so one day Steve Craken became Steve Germany and he had a sweet MSU ID that got him into the pool!

What the SCUBA Students weren't told is that they had to do the underwater skills portion of the class in Oxford MS over an hour and a half away. They were duped. So there was Steve at the dive shop in Oxford Mississippi, where they had a three foot deep by eight foot diameter kiddy pool to learn SCUBA diving. Not only was it ridiculously small but it was filthy which later gave Steve a sinus infection. You could not see underwater to the other side. The dive master said his pump wasn't working and it was just algae so everyone needed to get in or at most eight of the twenty -five students could get in.

Steve was in the second group to go and he already learned that in diving you always go with a buddy. They hadn't partnered up yet and the second group was assembling their regulators to their buoyancy control devices (BCD).

There he was "Mr. No Life Skills". Steve happened to be

right next to him. The Dive Master would give them an

instruction and show them how to set up their equipment and

this guy after every step was, **"How do you do that, How

does that go on?"** He was freaking out. Steve helped him the

first time but then saw the writing on the wall. *I have to

get away from this idiot. He is going to get someone

killed.* Thought Steve.

Steve grabbed his gear during a break and moved across

the room from "Mr. No Life Skills" to a dude that looked

calm and collected. Steve said, "I'm Steve do you have a

buddy yet?" and the dude was like, "I'm Phil no would you

like to buddy up?" Steve was elated to be away from "Mr. No

Life Skills" because he was having issues the whole class

and was popping up out of the shallow water every 2-3

minutes with a problem in the pool.

Sunday 0900 September 18, 1994
Ponce De Leon FL

Phil and Steve hit off the first day they met. Steve

told Phil the story of "Mr. No Life Skills" on the way down

to Florida. Phil said, "He will probably get someone killed." "That is exactly what I thought," exclaimed Steve.

They dove the springs at Defuniak and that was way cool. Crystal clear water in comparison to some 30 foot visibility in the Ocean that they dove yesterday. The owners also stocked the springs with some large fish. The previous day there were some small striper fish and you could see the fishing lures from the fishermen from the jetty.

Yesterday they did a shore dive and it was OK. There was not much sea life except for the small striper fish and...a single shrimp walking on the sandy bottom about 40 feet down. Steve swam down to check him out. When Steve got close to the shrimp, it jumped up about 3 inches off the sand, flicked his tail and jetted backwards at 40 miles per hour! It was about thirty feet away in 1.5 seconds. Steve had no idea they could do that. He then heard a clanking noise which meant the dive master wanted you to look at him. He instructed Steve to go to the surface. When they breached the dive master chewed out Steve for leaving his buddy to chase the shrimp. Steve was like ok fat dude got it don't leave my buddy.

Later that evening everyone was dead tired of two diving days. They went to dinner and Phil overheard the fat dive master say he wanted to keep his weight up so his wet suit would still fit him and Steve was like, "Are you serious?" Phil said, "He was!"

Steve and Phil were now certified open water divers!

12.

Saturday 1000 October 7, 1994
Drill weekend
Thompson Field ANG Base Jackson MS

"Take some of those Airmen and go clean out the test

cell. All the jets are deployed and we don't have any work

for them on engines today." Mumbled SMSgt Giles to SSgt

Rick Gernes. "That should keep them out of trouble." The

test cell was a building that jet engines were pulled into

and it had a fixture on the ceiling that mimicked the

airplane's wing pylon. The mechanics would install the

engine and then hoist it up into the fixture. Then they could run the engine after a rebuild and measure various aspects of temperature and vibration to ensure the engine checked out before releasing it back to the aircraft.

Rick came out of the bench stock room and the first Airmen he saw were Craken and smith. "You two get a fire hose and come with me." "Oooooh sweet a fire hose." SSgt Gernes told Airman Craken to drive. They all three climbed onto the yellow tug with the fire hose and cleaning supplies. The tug…was a small bright yellow vehicle designed for one person. It had been "Modified" by the engine shop with extra seats…and the welds showed it. Tugs are used to move jets and jet engines around the base (on trailers). This tug wasn't big enough to tow an aircraft but it was one of those type vehicles. The tug was meant to be driven slow and therefore had zero suspension. Steve was having a grand time seeing how fast he could get the tug to go on the open road between the jet engine shop and the test cell on the west side of the small base. There was a speed bump on the road, Steve saw it but did not know there was no suspension on the tug. The tug was a heavy vehicle with steel weights added so it could tow the heavy engines. Its top speed was not that fast but with no suspension the tug jolted violently over the speed bump and nearly launched SSgt Gernes off the tug. AMN Smith said, "GEEZ STEVE!" SSgt Gernes said after a dramatic pause, "You do that again and I'm going to revoke your base driving privileges."

The trio made it to the test cell without further incidence. They unpacked the fire hose and about that time TSgt McCormick drove up in the shop's blue F350. He was in charge of running the test cell. He started barking orders at the Airmen and then left. Steve and Jordan were playing with the fire hose when Steve had a great idea. The fire hose has a nozzle on the end to control the flow of water but there was also a valve on the wall that could shut-off the flow and it was behind the massive test cell door. So Steve and Jordan shut off the valve and rigged the fire hose up by weaving it through an A frame ladder. They tested it by turning the wall valve to full open. The hose waved up and down with the water pressure. They fed a little more slack through the ladder to get faster oscillations and aimed the ladder at the door where they thought The TSgt would come through. About ten minutes later they heard the truck and got ready. Someone came through the door and Steve opened the valve all the way.

SrA Lance was doused and ran out of the test cell! That was fine with Steve and Jordan because Brandon Lance was a little bit of a jerk.

By the time the test cell was "clean" there were two inches of standing water on the floor.

13.

Thursday May 13, 1994
Pelham Al. Blue Water Park

    Mat Hattier and Steve had just entered the cold water

of Pelham Blue Water Park. Mat was working on his SCUBA

Certification and asked Steve if he wanted to go. Steve had

never dove an old rock quarry so it interested him. Steve

got to do some diving around the quarry while Mat was doing

his skill checks. The visibility was about 40 feet in the

morning before all the students kicked up the silt on the

bottom. Mat got certified and they hung out in the quarry

for a little bit. Not much to see. Rock over there, silty

bottom, some other students doing stupid stuff. *What was that?* Steve asked himself. He dropped down 20 feet and saw several vehicles that were sunk in the quarry. There was even an old Robinson Helicopter. Never fly on a helicopter. They will kill you.

There were no hot girls but that didn't matter because both Steve and Mat were dating girls back in Starkville MS. Steve and Mat went to high school together and went on various adventures in rappelling including one high ropes rescue of a girl we shall call…Carla (because that was her name).

There they were on the cat walks of the Tennessee Tom Bigby Water Way Bridge up 70 feet rappelling off them into the water of the river and probably cutting class. Carla started rappelling down and got her hair tangled in the descender about half way down. Mat rappelled down on the rope across from her and was helping her. Steve saw what was going on and yelled up to cut her hair. Carla freaked out and yelled like she just had a limb severed,

"NOOOOOOOOOOOOO!"

Carla was gorgeous with long curly hair and she was in Steve's Spanish class. He asked her out and she accepted but then she said she couldn't because she didn't know she was still dating her x-boyfriend...WHAT...um OK.

So cutting her hair, the safe thing to do was out. Mat kept working to get her hair out and she started descending the rope and finally dropped free of the rope into the water. What they didn't know is that her hair was still in the descender and she couldn't swim because her head was attached to her waist and harness via her hair. So Mat jumped into the water on top of her and lifted her up so she wouldn't drown. Steve dove in and swam out to pull her in to shore. Steve was a boy scout and knew water rescue. He was swimming Carla back to shore and she kept saying, "Ow" because it must have been pulling her hair. Oh well another interesting day in Starkville MS. At least they weren't at school, they thought.

Then Steve was on the cat walk. He looked down and some kid was cooking his bacon that he brought for lunch! What a jerk.

14.

Friday March 12, 1999
Starkville, MS - Home of Mississippi State University - The
Bulldogs!

Steve Craken went in to the ROTC Detachment, Det. 425…
the best detachment in the whole Air Force ROTC. He is
days away from graduation and commissioning in the world's
best Air Force, The United States Air Force. He had a
rough couple of semesters leading up to this day. He had
spent five plus years learning to be an aerospace engineer.
Steve was fairly smart, as compared to his peers in high

school. He graduated in the top third of his class. He was definitely not a genius but he liked to tell new people he met that he was, in fact a genius… like Wiley Coyote.

Back in sixth grade at the end of the school year some teacher gave the class a math standardized test and she said, "Don't worry this test doesn't mean anything." So Steve decided to make some cool patterns on his Scantron card. Then a few days later it was SUMMER!

Steve's seventh grade year started out Ok except one day an office worker came to his pre-Algebra class and asked the teacher for Steve to go with her. So he picked up his books and followed the office worker to a special class room. He was in this "class" with another kid, Randy he knew to be not so bright. The teacher asked Steve and Randy to work on math problems with this math robot. It was a stupid toy designed for a six year old. You could change out cards to do basic math like 5 + 2 = 7. Steve finished the hour's work they had for him in 10 minutes. The other kid was struggling. Steve had time to ponder the great mysteries of the universe and finally asked himself a question. *"Steve, why did they pull you out of Pre-Algebra to do this second grade math with this moron Randy?"* OH NO!

THAT TEST AT THE END OF THE YEAR DID COUNT FOR SOMETHING! I HAVE TO GET OUT OF THIS SPECIAL CLASS!

What he lacked in brain power he made up in leadership skills and an attitude of never giving up. He was tenacious! Steve wanted to get a degree in aerospace engineering because he wanted to know everything there was about aircraft and flying. Steve developed a theory about engineers. Either you were a decent engineer and had people skills or you were a brilliant engineer but couldn't talk to anyone or convey your ideas. No people skills!

He was long past the "weeder" courses (difficult courses made to separate those students serious about engineering and those destined for the McCain School of Business) such as calculus and chemistry. His course load was fairly straight forward with no surprises and mostly just projects.

One rough spot was his grandmother passed away in Texas and he couldn't go to the funeral because of his courses. Another hardship was he had been awarded a navigator slot his junior year. A navigator is similar to a pilot and is also an officer. A navigator flies on airplanes and manages the weapon systems or defensive systems. Lt Craken lost his navigator slot due to a depth perception issue with his eyes. So there he was about to

commission in a few days and had no clue what he was going

to do in the Air Force. Another hardship was his girlfriend

of four years up and freaked out and wanted to break up.

All she said is that she didn't know what she wanted out of

life. Satan was really hitting him hard this semester. He

clinged to God and read the following verse every day:

Psalm 27:14 Wait for and confidently expect the Lord; be

strong and let your heart take courage; yes wait for and

confidently expect the Lord.

His boyhood dream was to fly but he was farsighted and

this was in the stone-age where you had to have perfect

vision to be a pilot in the Air Force. One of his fellow

cadets, Rob McAllister was in the office and joked with him

that he was going to be a base snacko in North Dakota. A

snacko officer (Official Title: Morale Welfare and

Recreation Manager) manages the bowling alley or a few

other similar type establishments on base.

The Det Non-Commissioned Officer (NCO), Technical

Sergeant (TSgt) Brown, was looking at Air Force jobs that

were still available this late in the assignment cycle.

There were not many. TSgt Brown said "You are going to

Tinker Air Force Base to be a structural engineer". Cadet

Craken said, "So, is that in Oklahoma?" TSgt Brown said,

"I think so." Good thing Steve paid attention and actually

liked his aircraft structures courses. MSU has a solid

engineering department that Steve found out later because

he would use a lot of what he learned for in his first Air

Force assignment.

Steve checked his watch and realized it was time for

him to meet his Structures Three class in the testing lab

of Walker Hall to test their mock up aircraft panels. There

were eight students that were seniors in the Aerospace

Program. Their last project before graduating was to

design, optimize and manufacture an aircraft fuselage

panel. The Students were assigned a team mate which made

four teams. Each team would produce two panels. Each team

were given four sheets of 2024-T-3 Aluminum. Each sheet

measured 2 X 3 feet. One sheet would act as the panel skin

and then the students had to shear another sheet and "break

form" the flat sheet into stringers. They had to design;

how many stringers they needed to attach to the panel, the

dimensions of each stringer, how to attach the stringers to

the panel, how many fasteners they would need and the

fastener pitch. Ben and Steve partnered up. They were

friends and worked on projects together before. Their

engineering chant was, "We suck" because aerospace is a

hard degree!

Ben and Steve decided on a hat cross sectional area

for the stringers and to go with a smaller leg dimensions

so they could use four stringers on the panel instead of

three. They also decided to support the flanges of the

stringer and install them with two legs attached to the

skin verses installing them through the center of the

stringer like conventional designs. They did some recon on

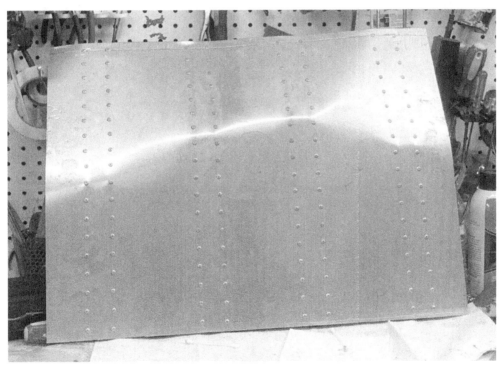

Figure 2 Manufactured Skin Panel Outer side

the other teams' designs and they were the only ones

installing the stringers or essentially, columns, by

attaching the legs on the skin panel. In effect this design

Figure 3 Manufactured Panel Inside aircraft view

simply supported the legs and allowed the column to support

more weight or stress without buckling or crippling. The

panel weighed approximately 2.1 pounds. The panels were

loaded in the hydraulic press like you see in Figure 3 with

the stringers taking the vertical load of the press' load.

Steve's panel was crushed after taking approximately 9,523

pounds force. Ben's panel withstood an amazing 11,252

pounds force!

The team assessed the difference in force before crushing/crippling columns in the two panels. They determined it was due to a manufacturing defect Steve had on his panel. When Steve was assembling his panel he was thinking of his girlfriend and why she was thinking of breaking up with him. He was distracted and miss-drilled a hole in a stringer too close to the stringer wall. Ben said, "That panel is yours." Steve answered Ben, "Ya OK." and finished his panel without further defects. The competition would not allow them to use another sheet and make a new stringer.

Some of the other teams used three really large stringers attached to the skin panel through the middle of the stringer with the legs facing away from the panel and free. When those panels were close to buckling the stringer legs started forming some wicked elastic hour glass shapes due to the pressure they were taking.

Ben's panel was able to withstand the most stress followed by Shane's, Steve's and then Christine's. The Professor said Shane and Christine's team won even though their panels didn't take the most stress because their design was easier to fabricate - less rivets and less stringers. There was no prize for the competition but Ben and Steve felt they were ripped off. No need to complain

because hey no prize and they were graduating soon.

15.

Saturday May 13, 1999
Starkville, MS Humphrey Coliseum

Graduation and Commissioning went well. It was a very

exciting and lonely time. All Steve's friends were

commissioning or going home for the summer since they were

not graduating yet. Steve was contemplating his

relationship with his X-girlfriend. Why did she break up

with him? Steve thought it was because she had always

lived in Mississippi and was scared to travel the country

and possibly the world. Her family came to commissioning

but she was nowhere to be found. He was really sad about

all of that but he trusted that God would get him through

that rocky time. His family lived a couple hours away in

Florence MS so he saw them at commissioning, graduation and

lunch after and then he was on his own. No

responsibilities. A valued moment in 2nd Lt Craken's life

was that his father, Major Craken (Army) was the officer

that commissioned him.

The lonely time didn't last long because Lt Craken had

a report date of June 1, 1999 to San Antonio TX for a month

long course and then to Tinker AFB OK. Since he was

commissioning in the Air Force he quit his job as the night

auditor at a hotel on the highway 25 bypass in Louisville,

Ms. He had no income but he did have a couple of credit

cards. His parents were awesome and let him have a gas

station credit card. You don't even know all the meals you

can make with gas station dry goods until you are homeless

and jobless. So he lived on credit about a month and

bunked with ROTC buddies during the week and hung out with

his x-girlfriends brother in Noxapater MS on the weekends.

He also bought a new Dodge Ram Quad cab that was red

because he had a family friend that knew he was

commissioning and would have income in the next month.

This was the nicest vehicle he had ever driven. His family

pretty much drove the wheels off their vehicles and got the

most use out of them.

His first car, his parents bought him for $500 cash.

It was a silver Pontiac Grand Prix, which sounds nice

except the year it was made was 1979. In 1979 Steve was 5

years old. Steve was super stoked and thankful to have his

own vehicle and did not have to share the baby blue ford

Escort Hatchback with Betty.

The car normally smoked out of the engine bay. One

summer day, Steve was a high school senior with zero cares

in the world. He had the sun roof out and windows down.

He pulled up to a traffic light and a car load of girls

pulled up next to him in a new convertible Mustang (with

the sweet 5.0 liter engine). He was like, "Hey how are you

girls?" The ugly girl in the back seat said, "Hey your car

is smoking. Is it on fire?" and they all giggled. Steve

just replied, "Nope that's normal" and dusted that Mustang

because that car had one heck of a V8 engine and four

barrel carb. It would haul and those girls didn't know how

to drive that Mustang!

16.

Monday May 31, 1999
Lackland AFB, San Antonio, Texas

San Antonio was fabulous! Free hotel room, free food
and making $24,000 a year in a beautiful city in Texas. Lt
Craken was no longer homeless or jobless! There were about
200 other brand new Lieutenants in the Acquisition 101
Class so there were plenty of people to make friends with
and there were also a couple of buddies from Starkville
that also had training in San Antonio.

Lt Craken knew from his days with the Mississippi Air
National Guard that he had to go to the Air Force finance

department the first day to get paid so he told his buddies
to go with him so they could get paid. While in line at
finance one of Steve's friends was talking about what the
go-to clubs were to find the ladies. He was concerned it
was going to be difficult to meet ladies. The finance NCO
said, "You won't have any trouble finding women because now
you are a walking benefits pole". Yes they were. In the
military they had free Healthcare, dental, life insurance,
housing and education.

     The course was not that difficult and it went well. It
was Acquisition 101. How to buy stuff in the Air Force. Not
like toothpaste or pizza but whole weapon systems like the
B-1 bomber. The class covered the "Cradle to Grave"
concept. Everyone in the course went their separate ways
after a month and Lt. Craken left for Oklahoma City.
Traffic leaving San Antonio on July 2nd was insane. It was
bumper to bumper 80 miles per hour on I-35. Steve
maneuvered into the fast lane between two trucks and stayed
there until well out of the city.

Part Two:

Arrival

17.

Wednesday July 5, 1999
Midwest City, OK

    Lt Craken reported for duty at the Oklahoma City Air

Logistics Center (OC-ALC) on Tinker AFB, Ok. When an

officer reports in for duty to his new commander it is

customary to report in wearing the service dress uniform,

which is the formal blue uniform. Since Lt Craken served

four years in the Air National Guard he had more ribbons

and medals than the typical Lt. He reported in to the

commander and the commander saw the ribbons and commented

saying, "It appears that you are not the typical stupid

second lieutenant". Steve accepted the compliment and explained that he used to be a jet engine mechanic in college.

Lt Craken started his job in the C/KC-135 Strato Tanker System Program Office (SPO). The SPO was responsible for managing everything about the massive jet. The 135 was not the biggest jet in the inventory but it was large. The tanker could carry up to 200,000 pounds of fuel, cargo, people or a combination. There were also other special purpose C/KC-135s. NASA owned one to train astronauts in zero gravity. The pilot would fly a series of large parabolas in the air. The airplane would ascend to altitude and then dive at a 60 degree down angle. At that point all the occupants would be subjected to zero gravity or 0G. At the determined lower altitude the jet would then climb at a 60 degree angle and all the occupants would be subjected to twice the amount of normal gravity or 2Gs. The jet had the nickname "The Vomit Comet".

Lt Craken's job title was structural engineer but they always put, "and other duties assigned" at the end of the orders to basically cover any job that needed to be accomplished. They would end up giving him other duties including the base Aircraft Crash Recover Engineer, Field

Call Engineer, Production Depot Maintenance Liaison

Engineer, Combined Federal Campaign Manager for the SPO

(the federal government's charity program), Sponsor for new

officers coming into the SPO, training interns and…

Aircraft Battle Damage Repair (ABDR)-by far his favorite

aspect of his new career.

18.

ABDR is an art. It was created after Vietnam when
U.S. Aircraft were being attrited due to battle damage and
they were parked on the ramp unusable for sorties. There is
also another organization close to ABDR called
Survivability/Vulnerability Information Analysis Center or
SURVIAC. SURVIAC analyzes battle damage reports on
specific airframes and combines them all together. If
SURVIAC has a report that means the airframe made it back
from battle. They provide a composite of that aircraft and
anywhere they do NOT have a record of battle damage that
means that either the aircraft had never been hit in that

location or more likely that if it was hit in that location
it did not make it back to base. There was such a
historical document on F-4 phantoms that served in the Air
Force, Navy and Marines during Vietnam. SURVIAC deduced
there was one location that the aircraft would not return
from battle if hit. This information was highly classified
so only a few engineers were privy to the information. F-4
Engineers discovered that the triple redundant hydraulic
system all came together in one specific location – where
SURVIAC concluded airframes never made it back if hit
there. The F-4 Program Office redesigned the hydraulic
system and removed the number 2 hydraulic system from that
junction and rerouted it. This redesign saved untold
pilots' lives. ABDR feeds damage reports to SURVIAC.

ABDR History. "During peacetime operations, damaged
aircraft are repaired in accordance with specific Air Force
technical orders, commonly referred to as the -3 (dash
three) series. These technical orders, generally written by
the airframe manufacturer and updated by Air Force Depot
Engineers, provide repair techniques intended to return the
aircraft to original design specifications. These
techniques require that repairs are performed to retain
design factors of safety, aerodynamic contours, and general

appearance. Typically these repairs are lengthy operations which demand numerous spare parts as well as specialized tools and equipment. The stress and pressure of combat operations usually preclude this type of repair-thus, the advent of ABDR.[1] The purpose of ABDR is to get the aircraft back in the fight. Historically, between three and five aircraft are damaged for every aircraft destroyed for tactical air missions.[3] Analyses of hypothetical future conflicts indicate that the ratio of damaged to destroyed aircraft could be as high as 15 or 20 to 1.[4] The Persian Gulf War clearly demonstrated this phenomenon. As many as 70 of the 144 A-10s deployed to the Persian Gulf were damaged, but only five were destroyed.[5] This large number of damaged aircraft highlights the need for an effective ABDR Program.[1] Also ABDR can be used to quickly return an aircraft to battle even in a degraded state where a permanent repair could take weeks to accomplish.

19.

Monday August 2, 1999
Battle Damage Lab Tinker AFB, OK

    Several lieutenants were assigned for training at the

Aircraft Battle Damage Repair lab on the south side of

Tinker AFB. The qualifications to be an ABDR Engineer are

a degree in engineering and be assigned to one of the three

Air Logistics Centers (ALC). Tinker AFB, Hill AFB and

Robins AFB are the ALC locations. All of the lieutenants

had degrees in aerospace or Mechanical engineering. There

were three males and two female officers attending training

including Lt Tiffany Jones. She was an attractive officer

and weighed in at probably 105 pounds. She was spunky too

and didn't take any ribbing from anyone. Lt Craken noticed

her fairly quickly but had a rule against dating officers.

If the relationship became serious and they got married the

Air Force had zero obligation to assign them together and

they could potentially be stationed thousands of miles

apart.

There were several ABDR classes to take before an

engineer was qualified and assigned to lead a team. There

were specific courses and general courses. The specific

courses were; Assessor, Engineer, Structural, Electrical

and Hydraulic. The generic courses were many but really

the only important ones were Small Arms (firearms)

Training, Survive and Operate, Chemical Warfare, and First

Aid. Most of the ABDR training was held a Tinker AFB at

the crash Lab. The first course the officers started was

the Assessor Course. This course was fairly short and

mostly in the class room. The course consisted of what

happens when an aircraft lands at the Forward Operating

Base or FOB with damage. Damage can be from battle or it

can be organic…ramp rash. Ramp rash happens when something

runs into an aircraft. This phenomenon is not restricted

to FOBs. There was an incidence in Florida where a young

Security Forces Airman drove his patrol car into an F-15.
Lt Craken found this course boring.

Typically on an ABDR team the enlisted team leader
would handle the assessor job and it was mostly learning
the Technical Orders or TOs. During a break the Lts were
in the break room and one of the Master Sergeants (MSgt)
came in and was talking to Lt Craken. MSgt Thomlinson was
bragging about his girlfriend who happened to be married to
someone else. Lt Craken ordered the MSgt to come into the
breakroom and sit down. Lt Jones came in also because she
didn't have much experience leading NCO or enlisted
personnel. Lt Craken said, "Stop telling me about your
married girlfriend because I don't want to have to testify
against you in a court martial, and also you could get shot
messing around like that." The Lt and MSgt were friends
and the MSgt had worked on the LT's truck so it was coming
across as a friend and not like some jerk officer. He said
"Ok" and left the breakroom. Lt. Jones quietly said, "You
are really good at that". He said, "Thanks" and they went
back to training.

MSgt Malcolm was giving an overview of the training
and touched on small arms training. He stated the officers
would be qualifying on the M9 Beretta 9mm handgun. Lt

Geronimo (Gerry) Ford asked MSgt Malcolm if, after the training, if they would be carrying the firearm around base. MSgt Malcolm thought he was kidding. Lt Craken knew he wasn't and was thinking to himself this does not help the "stupid Lt" stereotype. MSgt Malcom politely said, "No you only will be issued a 9mm if you deploy."

The next two weeks was the Structures Course. This was really a cool class because the students got to break a B-52 flap, which is about the size of a Cessna wing, with a crash ax. The flap on an aircraft serves as a mini-wing that changes the lift characteristics of the main wing. So to over simplify what a flap does, when the flaps are deployed or extended they allow the aircraft to fly at slower speeds and the flap plus the wing create more lift. The reason you want to fly slower with more lift is for take-off and landing – so you don't need 4 miles of runway to take off or land. After take-off the flaps are retracted, which makes the wing have less drag and the airplane flies faster.

The instructor MSgt Malcolm was a good instructor and knew his material. MSgt Malcolm lead the officers into the lab area and went over safety (safety first). So Lt

Craken, Lt Tiffany Jones, Lt Gerry Ford, Lt Mike Carry and Lt Sarah Timmons went into the lab.

MSgt Malcolm assigned workbenches to the engineer/students. He instructed them to open their binders and start project number one. Project number one was to cut a push/pull rod into two pieces and then apply a repair to it using materials found on the ABDR trailer.

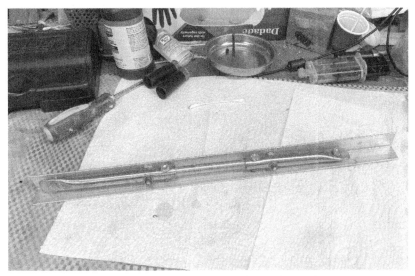

Figure 4 Push Pull Rod in its Jig

Push/pull rods are common used on old aircraft to control engine functions and flight controls so the chances that one of these rods could become damaged is fairly high on an old aircraft such as the B-52 or C/KC-135.

Most everyone finished the project that day except for Lt Ford. He needed some assistance.

The following days the engineers worked on a few other projects like making a pulley wheel and a simulated skin panel damage repair with laying out a fastener pattern. Once those projects were complete it was on to the Final project. This project would take the better part of a week.

The students arrived in the lab and MSgt Malcolm asked if anyone wanted to use the ax. Lt Jones spoke up first and grabbed the ax. MSgt Malcolm drew a circle on the B-52 flap where he wanted the damage inflicted. Lt Jones probably never wielded an ax before this day and gingerly hit the aluminum skin on the flap. The ax knocked some paint off and maybe dented it a little. Then Lt Craken said, "Can I try?" Lt Jones gave the ax to him. Lt Craken lived in that old house in Mississippi that was part of a plantation at some point in the past. Inside the house in the summer it was hot and in the winter it was cold. The house had zero insulation in it and routinely had a family of squires living in the walls. The only source of heat for the living room and kitchen was a fire place so Steve was well versed with an ax.

He was also trying to impress everyone so he really put some muscle into the swing and hit and damaged the flap well outside of the mark the MSgt made. The MSgt said,

"**Whoa**", so Lt Craken said, "No, no, I've got this". So he aimed better and hit within the circle and tied in the previous damage. Then another Lt wanted to give it a try. Then Lt Ford took the ax. Lt Ford was from backwoods Nebraska (If Nebraska has woods). Lt Ford was fairly tall at 5'11" and a fit 200 pounds. Conversely Lt Craken was short at 5'6", 150 pounds but also fit. MSgt Malcom told Lt Ford to flip the ax over from the curved blade side to the spike side to then swing inside the flap and damage the internal rib. So MSgt Malcolm drew a circle inside the flap on the rib. Lt Ford was also trying to impress everyone, including Lt Jones and he ended up swinging wide and missing the mark really wide. MSgt Malcom exclaimed, "**WHOA**", and grabbed the ax and put it away. He should have known not to give lieutenants a crash ax or really any tool.

With the last swing of the crash ax the fun was over. It was time to dive into the tech orders and find out information on what type of materials and thickness were damaged by the overzealous ax wielding male Lts.

When Boeing designed the B-52 they devised a way to know where something is located on or in the aircraft. A map of the aircraft structure. Typically the datum plane or starting point for Beam Body Stations (BBS) started at zero on the nose of the aircraft and proceeded aft in inches, (usually called the "X" direction). See Figure for a

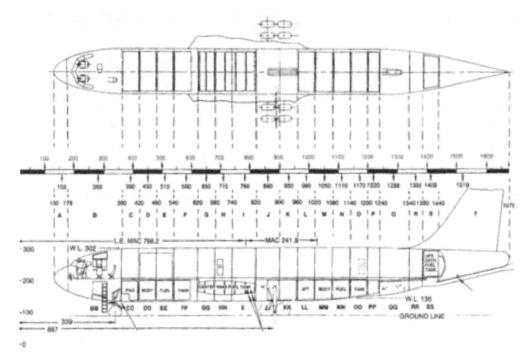

Figure 5. C/KC-135 Structure

pictorial representation of BBS for the C/KC-135. For example if I was looking for the 820 bulkhead on an aircraft it would be BS820 inches aft of the nose. For the C/KC-135 this is not true because the datum plane is a few feet forward of the conventional nose due to the longer

"Snoopy nose" on some 135 variant aircraft. That aircraft

had a longer nose storing some classified equipment in it.

The manufacture also devised location systems for the

"Y" and "Z" directions. The "Z" direction or up is called

Water Line or WL. These systems are derived from ship

building. The left and right direction "Y" is called the

Beam Butt Line or BBL. The left side of the aircraft is a

positive number and the right is a negative number.

When it comes to the wings the measuring starts at the

center of the fuselage and moves outward toward the wing

tip. The measurement is denoted as Wing Station (WS).

There is a right wing station and a left wing station. So

a wing rib on the right wing at 500 inches would be denoted

as RWS 500.

The flap was oriented in the same way but with Flap

Station or FS and if the aircraft had more than one flap

per wing it would also have a number starting from left to

right as if you were looking from behind the aircraft.

The B-52 has four flaps and the one in the shop that we

were wood chopping on happened to be the number one flap

(from the left wing). It is the smaller flap but still the

size of a Cessna 152 wing.

MSgt Malcom brought out the ABDR B-52 Technical Orders

(T.O.) 1-1H-39-B52, which contain all the relevant data to repair battle damage on the B-52. The Figure shows the B-52 without skin so you can see the structure.

Figure 6. B-52 Structure

All the engineers had their own T.O. to look up the area in question. Lt Craken opened the T.O. to the #1 flap page and began looking for the damaged area. The damage was located on the skin on the number six rib. The rib was also damaged along with the rib web. The rib is basically an "I" beam that is curved in the shape of the flap.

Once all the original materials were known it was time to design the repair with materials that were commonly stocked on the ABDR trailer that would accompany the team to the forward deployed location.

This course was written for the technicians so there was no need for any strength or moments of inertia calculations because everything was factored with a healthy safety margin. Also the repair tech order assumed the same material as the original damaged material and one gauge thicker so the repair was actually stronger than the original structure. See Repair Notes at the end of the chapter.

1. ORGANIZATION	2. LOCATION	3. MDS	4. SERIAL NUMBER	5. DATE	6. DAMAGE NUMBER
507 CLSS	Tinker Lab	B-52 Lab Flap	60-033		1, 2, 3

Repair

Damage

Damage #1

Clean Damage

Side View Inside Damage

Damage #3

Rib cap

Web Repair

L" Angles

Web - Damage #2

Damage #2 Original Material of web 0.040" Thick Al 202A-T351. Repair Material 0:053" Thick 2024-T3 CIAD

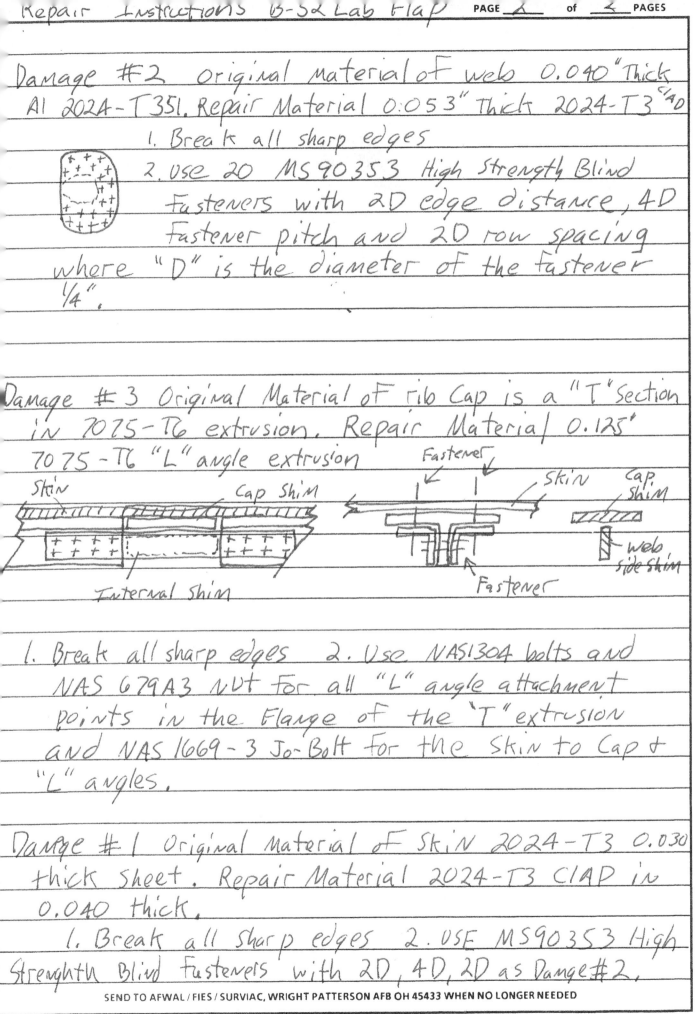

   1. Break all sharp edges

   2. Use 20 MS90353 High Strength Blind Fasteners with 2D edge distance, 4D Fastener pitch and 2D row spacing where "D" is the diameter of the fastener ¼".

Damage #3 Original Material of rib Cap is a "T" section in 7075-T6 extrusion. Repair Material 0.125" 7075-T6 "L" angle extrusion

1. Break all sharp edges 2. Use NAS1304 bolts and NAS 679A3 Nut for all "L" angle attachment points in the Flange of the "T" extrusion and NAS 1669-3 Jo-Bolt for the Skin to Cap & "L" angles.

Damage #1 Original Material of Skin 2024-T3 0.030 thick sheet. Repair Material 2024-T3 CIAP in 0.040 thick.

   1. Break all sharp edges 2. USE MS90353 High Strength Blind Fasteners with 2D, 4D, 2D as Damge#2.

☆U.S. GOVERNMENT PRINTING OFFICE:1989-242-893:75064

20.

Saturday August 14, 1999
West Pool Huntington Place Apartment, Midwest City Ok

Steve woke up in his one bedroom apartment and asked out loud to himself, "Steve what are we going to do today?" He thought, *Maybe I'll go to the hardware store and buy some metal to make a coffee table using ABDR techniques. First though I'll make some blueberry crumbly muffins and take them to the pool.*

Steve baked some crumbly muffins and went to the pool. He thought, *Oh score. There is that cute blond girl he had seen walking around the complex. I probably should not date*

*her because she lives in my apartments. Maybe I'll say hi
and see where it goes. She looks really young I'd better
see if she has a job or is in college or high school. No
need to go to jail this early in my career.*

Steve said, "Hi, I am Steve. What is your name?" She
replied, "I'm Jewel. I live a few buildings over there."
And she pointed to the front of the apartment complex.
Steve pointed to his apartment that could be seen from the
pool and said he lived in that unit. They hit it off. Steve
found out that she went into the Air Force out of high
school and was a security Forces specialist for a while but
recently got out after only a year and a half. That factoid
was the first warning bell that went off for Steve. A tour
of duty is usually three years minimum so for her to have
gotten out early meant she was medically disqualified or
forced to separate for other reasons. Steve would have to
ask about that when he knew her a little better.

Steve, against his better judgement, asked Jewel out
for that night. There weren't any good movies out so Steve
suggested they rent a movie and he could cook for her. He
asked her if she liked steak and she did. Steve decided to
grill for dinner. He made some New York Strip Steak and
baked potatoes.

They had a fun date and watched a comedy movie. Jewel
said, "I'm going to be respectable and leave for the
night." Steve said, "That is a good idea. Can I walk you to
your apartment?" She agreed. They said good night at her
apartment and Steve walked back to his.

Sunday August 15, 1999
West Pool Huntington Place Apartment, Midwest City Ok

Steve headed to the pool after church and lunch. He
was hoping to see Jewel at the pool. She was there! They
swam for a while and then laid in the sun to get warm. They
made a day of it. Jewel said, I'm hungry but I don't know
what I want. Steve said how about we go to your place and
have dinner there. She agreed and they met at her apartment
after Steve changed at his place.

When Steve entered her apartment it looked like a frat
house. His apartment wasn't home to a neat freak but it was
clean and fairly de-cluttered. In jewel's apartment there
were clothes all over the place. On the floor, on the
couch. On the kitchen table. Jewel acted like living this
way was perfectly normal. Warning bell number two for Steve
but what could he do right now? So he decided to stay for

dinner. Jewel looked at Steve and said, "I can't cook. It is not a skill my Mom taught me or really I didn't want to learn." Steve replied, "No problem. I can cook. My Aunt gave me a book before I left Mississippi titled, A Man, A Can and A Plan. Basically you use a cream of something soup, add some cheese, some meat and some noodles and you got dinner." Let's see what you have in the pantry. Steve saw some tomato sauce and noodles. He looked in the freezer and found some hamburger and put it together with some spices to make some sphegetti.

Jewel and Steve liked each other. Unfortunately Jewel was an Airman Basic when she was in the Air Force and Officers were something to be revered from that entry rank. She kind of looked up to Steve more than she should have. They dated a short while and then she started smothering Steve. Steve went to church and afterword hung out at his friend's house. Then he went to small group and ignored the 15 calls from Jewel. He needed a break. He enjoyed the group tonight talking about picking up your cross daily and following Jesus. He had this nagging feeling subconsciously that he needed to break it off with Jewel. This was not a healthy relationship.

He drove home to his apartment. It was about 10 P.M. on a random Sunday night. He was thinking it was a nice break from Jewel. He would call her tomorrow and see how her Sunday was. He went into his apartment and got ready for bed and then climbed into bed. His bed was right next to the window and he saw the shadow of a person run by the window and heard them. He started thinking, *Oh no it's Jewel going to see if my truck is in the parking lot! I have a stalker! That is so cool.*

KNOCK, KNOCK, KNOCK…Oh no it was her. Steve groaned as he got up and got dressed. He looked out the window to see who it was. There was Jewel looking hot like she was ready to go clubbing. Steve turned the light on and opened the door. He said, "Hey Jewel how are you?" Jewel said, in an annoyed tone, "Where have you been all day?" Steve nonchalantly replied, "Oh I had church and then I hung out at Jerry's house until small group." She replied, "Well can we watch some TV?" Steve said "Ah, um I have to get up early for work and really I just went to bed." Jewel looked dejected so Steve added, "I'll call you at work tomorrow." She said, "OK." And went to her apartment.

Monday August 23, 1999
Cube Farm Tinker AFB OK.

Steve picked up the phone and called the local temp

agency that Jewel worked at and some irate sounding dude

answered the phone. Steve asked for Jewel. Irate dude said,

"Who are you?" Steve said, "Her Boyfriend." The irate dude

continued, "Where is Jewel? She has the keys to the file

cabinet and the server room." Steve replied, "What, she

isn't at work?" "NO and I have called her apartment and she

is not answering." Steve said, "I don't know gotta go" and

hung up. *What the hec? What is going on? Oh no she has a*

spare key to my apartment in case I locked myself out, he

thought. He didn't have a land line to his apartment. He

ditched those when he went active duty. He looked over at

Joe, his co-worker, who was grinning and staring at him

because he was living vicarious through Steve. Steve said,

"I'm going home for lunch." Joe responded, "Sounds like a

good idea."

Steve was imagining all kinds of scenarios like an

empty apartment when he got home. He opened up his

apartment cautiously looking for any signs of strange people or things out of place. Nothing. Everything was there including all of his firearms so he locked up and walked over to Jewel's apartment and noticed the eviction letter nailed to the wall next to her door. He knocked and Jewel opened the door and gave him a big smile, "Hey I missed you!" Alarm bell like number 20. Steve came in and told her about the conversation he had with her x-boss. She did not seem too concerned. Steve had enough warning bells and broke up with her right there. Too unstable. The relationship was a wild ride and that is what Steve wanted…he thought at the beginning of it. Just then Steve heard Jesus whisper to him, "This is not what I want for you. Be patient and I will show you what you need. Be kind to this lost soul."

Steve asked for his key back and then started helping Jewel make some good life decisions. A Couple of days later she knocked on his door and told him she was moving to California. Steve helped her pack and took her to lunch for the last time. They departed on good terms. Steve let out a big breath and was glad he didn't do something really stupid.

Joe said, "What happened to your stalker girlfriend?"
"She moved to California", said Steve with a relieved look
on his face. At that Joe nodded his head and turned around
and went back to work.

21.

Tuesday September 21, 1999
Yulin Naval Base (People's Republic of China PRC), Hainan
Island, China

    Everything in China is a cheap knock-off. They

haven't been great since they built the great wall or the

times of Genghis Khan. Buildings built in modern China are

built by the lowest bidder. The builders have to cut

corners or they don't make a profit and there are several

instances of brand new buildings that look great

architecturally from the outside but are not structurally

sound. They are not sound due to inferior concrete and

steel. China is a sham.

Captain Wong was nervous this early morning at 0300 (3 A.M.) as he inspected the submarine pins at the secret underground base. Everything was shipshape but the big brass were coming all the way from Beijing to oversee the transfer of special hardware flown straight from Mother Russia. Land attack missiles.

The special sub launched ground attack missiles routinely nicknamed "Vampires", by the U.S. Navy would definitely upgrade the Chinese nuclear subs. The Russian missile was code named the "Sampson" by North Atlantic Treaty Organization (NATO) but the export version, which was dumbed down a little, was the version being delivered today, for a high price to the PRC. The missile did not have a NATO code name because NATO didn't know the Chinese had them. China called them code name Dongfeng Fengbao or Fire Storm.

China currently had two super quiet nuclear fast attack subs. NATO and the U.S. Navy had zero intel of the two. The U.S. Navy kept tabs and general location of all enemy subs with our own subs SONAR and Low Earth Orbiting satellites.

The U.S. Air Force launched a classified number of sub detecting satellites that used magnetometers onboard to detect submarines made of Ferris metals. These satellites could not detect Russia's Typhoon class because their hulls were made of titanium. Thankfully China's were made of steel.

China's Communist leadership was looking for a way to show their dominance in the world and show up the U.S. Unfortunately for China the U.S. military was far more advanced than theirs until now, they thought. The addition of the unknown relatively new submarines that China bought from Russia a year ago partnered with the cruise missiles would put China on top and they would get to bloody the U.S.s proverbial nose. The exact time of the attack was unknown because China would need to pick the perfect time for the initial attack. This operation was known to five men of the Chinese politburo and they were sworn to death to not tell another soul. Operation FIRE LOONG or DRAGON FIRE in English was the code name.

The truck arrived and the offloading of the missiles began. Russia sent some technicians to help with the offloading and integration to the subs but China was not having any of it because they didn't trust their communist big brother so the Chinese military police sent them packing for home…rudely.

The dock workers inside the cave-like base were not especially skilled even by Chinese standards. They had no clue how to attach the special slings to the missiles so they did their best. When they lifted the first missile out of the truck the sling held for a few seconds but the dock workers only used two of the ten bolts necessary to carry the load of the missile. The left side bolt sheared first and then overloaded the right bolt and it sheared. The missile swung down violently into a rope handler and knocked him into the water next to the sub. The missile hit the concrete and the missile fuel tank buckled and let out the hydrazine that was under pressure. It did not catch on fire but anyone within 30 feet immediately died that inhaled the poison vapor. Capt Wong immediately hit the alarm and everyone not within 30 feet ran for the door. A special unit was called in to evacuate the gas and clean up the toxic scene.

Maj Huin, which was Capt Wong's immediate supervisor and responsible for the underground base was immediately taken to a "re-education" camp and never seen or heard from again. General Zia immediately promoted Captain Wong to Major and head of the sub base.

22.

Date 0800 Friday November 12, 1999
C/KC-135 SPO Commander's Office, Tinker AFB OK

Col. Baxter called in Lt Craken and his civilian supervisor Wayne Paxton to his office which was definitely abnormal. While walking through the cube farm 10 minutes ago the Lt asked if had done something wrong and his Boss just said, "Maybe". The Lt was thinking-great on a Friday no less.

The Col asked the men into his office but did not ask them to sit down. The Col began talking about a ground mishap in the millions of dollars range that happened back in September 1998 to a C/KC-135. This particular jet was

owned by the Mississippi Air National Guard out of Meridian

Ms.

Lt Craken had driven by the Meridian Regional Airport

several times before he went active duty and seen the C/KC-

135s on the ramp. Typically the jets that Air National

Guard units have they keep for the life cycle of that jet

so they take pride in ownership and those jets are some of

the best maintained aircraft in the U.S.

The tail number of this jet was 57-1418. The first

two digits of the tail number are the year it rolled off

the production line. So this jet first flew in 1957!

Col Baxter explained, "This jet was finished with the depot

process and had one final test before being flown back to

Mississippi and returned to duty. The test was a proof

pressurization test." A proof pressurization test is

designed to make sure the fuselage of the aircraft could

maintain an equivalent pressure inside of 8 Pounds per

square inch (psi) differential. What that means is sea

level is around 14 psi so it makes everyone inside the jet

feel like they are at a lower altitude than 40,000 feet

which if a person was subjected to atmospheric pressure at

40,000 feet (2.7 psi) they would pass out due to lack of oxygen.

The technicians performing the test were not following the correct procedure as laid out in the Tech Order. They made three critical errors. First they were using a homemade pressure gauge that did not have a needle stop installed, second they were using the massive engines instead of a ground air cart and lastly they had the emergency outflow valves capped. The outflow valves were designed to relieve pressure if it went above 9.5 psi.

The airplane had the new efficient engines, CFM-56 so it did not take long to pressurize the aircraft. The Technicians did not notice the needle on the gauge go completely around and they thought they were well below the 8 psi they were testing to.

While the commander droned on, the Lt was reminded of his first car (not the blue Ford, he shared that with the family). The 1979 Grand Prix with a four barrel carburetor sitting on top of 8 cylinders. That car was smooth and would haul butt. The speedometer did not have a needle stop and it went up to 85 MPH but the needle would keep going around the gauge and the fastest he got it to was 15

MPH. He estimated it was over 100 MPH on South Montgomery
in Starkville!

The commander asked the Lt if he was following and Lt
Craken asked, "What happened to the aircraft? Did it have
to go back through depot?" He said, "No the pressure blew
out the back bulkhead and the tail fell to the ground.
Boeing estimated the pressure went up to as much as 20 psi
and it was designed for something like 9 psi max." He
continued, "That aircraft was towed to Hanger 7 and
investigated, the hangar doors were welded shut and now it
is clear to do whatever the hell we can with it, and when I
say we I really mean you! We need that hangar space so go
get that wreck out of the hangar and park it at the compass
rose. Get the good parts off of it and back in supply. Cut
up the rest and get rid of it." Lt Craken said, "Ok, Yes
sir." And he left the commander's office relieved to not be
in trouble but also wondering how in the hell he was going
to do all that stuff he just said he would do.

23.

Wednesday December 29, 1999
Empire State Building, 20 West 34 St NY, NY

Jerry was a cool dude that Steve met a church. He was a native Oakie but was still cool. Steve knew Jerry a total of two months when Jerry invited him to go to New York City Times Square for the 1999-2000 New Year. *How cool would that be?* Steve Thought and said, "SIGN ME UP!" Jerry was finishing up dental school and going into the Air Force when he graduated. He was to become a member of…The Med Group. The Med Group encompassed all medical care for the services. They were Air Force but barely. To say they had their own way of doing things was an understatement. Jerry told Steve that his friends in medical school were also going on the trip. They weren't going into the Air Force when they graduated. Adam introduced himself at church. He said, "Hi my name is Adam and I'm a doctor, did I tell you I am a doctor?" Steve replied, "That's cool." Ronda was standing next to Adam and Steve introduced himself to her. She said, "I'm Ronda, no "H" and I work with babies."

They met each other two weeks ago and here they were on top of the Empire State building freezing their butts off. They came straight from the airport and hadn't had time to change so the temperature was 32 degrees and a thirty MPH wind at the top of 102 floors. The view was amazing!

They left and grabbed a cab to their hotel. Steve rode up front because they filled the cab up. The cabby whom was originally from some other country asked their group why they had to come to NY. They told him for the New Year's Eve Celebration. The cabby asked them if they came to blow up the city. They said, "WHAT? NO!" The cabby also told them they had been preparing for new Years for six months,

welding man-hole covers down, hiring extra police and Port
Authority. They were expecting four million people at Times
Square on New Year's Eve.

They arrived at their $200 a night hotel and it was
not anything special but that was OK because they were
there only to sleep and wouldn't even be there hardly New
Year's Eve.

They all got ready for bed around 11 P.M. because they
wanted to get up relatively early and go see as many sights
as possible the next day. Jerry took a shower but instead
of coming out wrapped in a towel he tried wrapping a hand
towel around his body. Steve Said, "Don't look Ronda!"
everything was hanging out! Jerry ran into the small closet
to get dressed but it was so small the door wouldn't close
with him in there and even if it did it was pitch black.
The hotel building was so old they had a steam heat
radiator and it was fairly cold in the room. The radiator
made weird clunking noises that woke Steve up a short time
later. When he woke up, he saw Adam flying through the air
toward him. Steve was sleeping on the bed next to the
radiator. Adam flew by Steve and was beating the radiator
with a shoe to make it stop clanking. Steve thought to
himself, *who are these Oakies?*

Thursday December 30, 1999
NY, NY

They left the hotel going to the World Trade Center
then Statue of Liberty, Wall Street, Rockefeller Center and
finish up with an evening at a fancy French restaurant and
then catch a Play, Cats.

Steve had been to NY when he was in Jr High on a band
trip so he had seen most of the sights but not the World
Trade Center. They arrived at the WTC around 1000. There
was a huge line going in to go up to the observation deck
so they decided not to waste a lot of time on that. They
did lay down at the base and look up the side of the
building to the top. It was crazy how tall that building
was.

The Statue of Liberty was awesome as usual. Jerry had
a phobia of being attacked by birds. Well where are most of
the pigeons in New York…At the statue of Liberty. He was
freaking out pulling his friends over using them as human
shields so he always had a buffer between him and the
flock. Steve was daring him to run into the flock on the
ground for $20 and he wouldn't do it.

They wound down the day by going to the fancy French restaurant Le Baudette on 5th Avenue. Now Steve had a very low opinion of the French. Ask any Frenchman what army marched under the Arc De Triumph first. If they will tell you, it was the German Army! Well this restaurant did not change Steve's opinion. He looked at the menu and he went with an expensive duck steak. Jerry ordered a fancy goat cheese salad.

When the food came out Steve had five duck steaks a little bigger than a quarter each. He thought, *I'm going to have to get some street pizza.* Then everyone looked at Jerry's "Salad". It was literally a piece of lettuce with a small block of goat cheese on it the size of a roll of quarters. Steve gave Jerry a duck steak and said, "Bon Appetite".

On the walk to Broadway Steve snagged a huge slice of pizza and was no longer disgruntled. He was very gruntled with the pizza.

One of the things on Ronda's list was to see a play and the one that fit their schedule was Cats. The men were less than stoked to watch grown people parade around like cats. Steve never like cats especially Butterscotch that was fond of pissing in his jeep. They went anyway and had some good balcony seats right next to the rail. During one

of the scenes a female "Cat" was crawling on the rail where

they were seated. She came right up to Adam and planted a

big wet kiss on his face. Jerry snapped a pic with one of

those disposable cameras.

Friday 1500 December 31, 1999
Times Square

New Year's Eve in New York New York! It was going to

be a fun day. The group did some more touristy things but

wanted to get to Times Square to get close to the concerts

and the ball. They left Central park and walked up 7th

Street bought some pizza and made it as far as 54 Street.

The police had this New Year's thing down pat. As soon

as a crowd came up they would block off the cross street

and the sidewalks. If you left after they blocked

everything off, the police would make you leave and go

further back toward Central Park away from Times Square. Ok

so no leaving you better stop drinking anything. They had

nine hours until the ball drop. The good thing about being

packed into the space like cattle was they were warm. The

temperature was hovering around freezing but with that many

people there was no wind. There were some cool concerts and

the people watching was awesome for the Oakies and Steve

from Mississippi.

Saturday January 1, 2000
NY, NY

Y2K. Year 2000. Everyone was mildly freaking out about

Y2K. No one knew what would happen because the coders back

in the day couldn't see that eventually they would need

four characters instead of two to define a year when the 20

century wound down. So in old code if you put in "00" then

the code would interpret that as 1900. Banks, navigation,

Wall Street stores all used old code so there was this mad

rush to hire coders to install year patches. Were ATMs

going to work? Were you going to lose your money? No one

knew. So the first stop after finding a bathroom was a stop

at the ATM. Steve put his card in and worked with the logic

and withdrew $100 to travel home today. It worked like

normal. They were all flying out today because no one flew

on New Year's Day. The flights were cheap and the airports

disserted.

Date 0830 Monday November 15, 1999
Hangar 7, Tinker AFB OK

    Lt Craken walked the ¾ a mile from his cube to Hangar

7. He found a door with a tiny window in it and being short

he had to jump up to look through it. What he saw was a

C/KC-135 jet with no tail. Ok found it. Now he pulled on

the door and it didn't budge. He thought it was locked but

looked closer to see it was still welded shut even though

the accident investigation was completed. He walked around

and found some offices and asked for access to the hangar.

The supervisor didn't seem too thrilled to help until he

found out the Lt was going to move the wreckage out. He

said it would be open by the end of the day.

Date 0830 Tuesday November 16, 1999
Hangar 7, Tinker AFB OK

    Lt Craken had told all of his Lt buddies about the

huge project and they wanted to see the wreck for

themselves so he escorted everyone over to Hangar 7 because

it was on the flight line and not everyone had a line

badge. He said, "I'll show you a picture on my computer of

what the airframe looked like just after the mishap, when
we get back to my cube."

Figure 7. Aircraft 57-1418 after mishap, Right Side.

Figure 8. Aircraft 57-1418 after mishap, Left Side.

The weld holding the door closed was ground off and
the Lt turned the lights on and went into the hangar. It
was a little eerie with the lights just coming on. At
least no one was really hurt in the mishap. One of the
technicians received a broken toe from the rudder pedal
suddenly moving due to the aft bulkhead explosion. Lt
Craken opened the crew entry door and installed the ladder
up into the cockpit. He went up followed by four other
Lieutenants. They all asked Lt Craken what happened so he
told them the story. This was going to be a massive

undertaking! This was day one of aircraft 57-1418 salvage

operation that would eventually take eight months.

After the field trip across the ramp, Lt Craken

settled in his cube and showed the other Lts the pictures

just after the mishap. The other Lts said, "Good luck with

that," and left him alone to do the mundane work of

starting a list of things to accomplish to officially

retire this C/KC-135

There were no aircraft mechanics in the program

office. All of the mechanics and support back shops were

located down on the line and under other offices and

commanders. What that meant for the Lt was he was going to

have to beg borrow and steal at least figuratively (or not)

to get this airframe disassembled and all the good parts

back into supply. There was an old crusty engineer, Walter

that saw the Lt struggling and so he came and told him

there were two other airframes retired for various reasons

over the years and the 654 Combat Logistics Support

Squadron or CLSS, provided the manpower for disassembly of

parts and finally they cut up the airframe. The Lt was

stoked because he had been attached to the 654 CLSS for

battle damage exercises, he had friends over there and he

was on tap for potential deployments with them.

The first and most urgent problem to solve was how to move the airframe. Hangar space is a premium especially because the alternative is outside on the ramp in the never ending Oklahoma wind. The commander told the Lt, he had one day to get the carcass out of his hangar. So the Lt went to the shop that tows aircraft and started asking questions. They said they would need an accurate weight and balance to be able to tow it. The Lt was working on his Private Pilot's License so he understood weight and balance…for a Cessna 150. How in the hell was he going to get the weight and balance for an airplane that was missing the aft 20 feet of fuselage, vertical tail and two horizontal tails?

He had a good idea, go to the crusty engineer. He went and found Walter eating a peanut butter and jelly sandwich in his cube at 1000 hours. Walter said, "Why wait till noon when you are hungry now?" The Lt said, "That makes sense." Lt Craken then said, "I need the weight and balance of 1418. How can I easily figure that?" Walter pulled the Structural repair Tech Order out, which was about six inches thick, and turned to the front where it laid out all the dimensions of the plane and weights for each large part like an engine or tail stabilizer, etc.

Walter said, "you can go through the tech order and subtract off all the parts that are gone and use the formula that is on page 1-24, and spend a couple of days or go ask Sheldon for his excel spreadsheet and get the answer in 3 minutes." The Lt had a healthy dose of procrastination and laziness so he went and got the spreadsheet… and it was a thing of beauty! All he had to do was subtract off the horizontal tail stabilizers and vertical tail along with an estimate for the rear structure that was missing.

Date 1645 Tuesday November 16, 1999
Cube Farm, Tinker AFB OK

With the weight and balance in hand he went to the tow shop and asked if they could tow it that day. The scheduler looked at the schedule and said they could possibly tow it on swing shift. The Lt was happy to make his deadline but not happy about working late. He had some church friends coming over for dinner that night. He asked the scheduler to call him before they towed it so he could provide support if needed. She said fine and the Lt went back to his office to get a game plan together for removing good and bad parts.

Steve had grown up in the Church of God because his Dad was a minister and Chaplain. He told him about a church in OKC that he knew the pastor and worship leader. So Steve checked out Crossings Community Church up north OKC. It had what he was looking for…A good singles ministry.

He had some friends there that were supposed to come over for dinner tonight. Steve had planned on making some shrimp fettuccini with salad and bread but it didn't look like that was going to happen. He called one of his friends, who also lived in the Village, and told her he was going to be late due to work and there was a key underneath a rock and to let herself and the rest of his friends inside.

Lt Craken called the scheduler shop and asked when they thought they could tow the plane and the scheduler said they are towing it now. *Wow thanks for the call*, he thought. So the Lt took off at a sprint to catch the tow crew ¾ mile away.

He caught them as they were towing the gangly bird out of the hangar and flagged them down. They said jump in the tug so he did. Apparently they didn't really need a weight and balance sheet after all. Normally the C/KC-135 is tail

heavy without fuel on board so it is equipped with a tail
jack if the ground crew had to empty the tanks. Installing
a tail jack every time the 135 was moved was really not
practical at depot because they are defueled as soon as
they come in so their solution is a 500 pound "pet rock" or
a big concreted block they place in the cargo area up
front. The pet rock ensures the jet does not perform a tail
stand and damage the tail boom.

Steve got a ride to his truck thankfully because the
compass rose was two miles away from where he parked. Then
he drove the thirty minutes from Tinker to the Village to
his friends waiting for him. All the ladies, who were just
friends, were digging the uniform because they had never
seen Steve in uniform. They all decided to order pizza.

Date 0730 Wednesday November 17, 1999
Cube Farm, Tinker AFB OK

It was time to devise a plan to pull off all the good
parts from the aircraft. The easier problem was to figure
out what the bad parts were. Lt Craken started thinking
about the over pressurization but then got side tracked.
Start with the big pieces. The four CFM 56 engines were

good. No damage to them. Those could be pulled off but
they would have to go back through the depot process
because they were sitting in a hangar not prepared for long
term storage so they could have serious corrosion on the
blades or combustion cans. Lt Craken called MSgt Morris at
the 654 CLSS and told him he could start removing items
from 1418. The landing gear shop called fifteen minutes
later and requested the landing gear after being cleared by
engineering because the shop were short landing gear to
move aircraft around the ramp throughout the PDM process.

When a jet comes into PDM the gear are removed because
the gear have to be overhauled as well as the airframe.
Then you get into a scheduling carnival game. An Airframe
is ready to be moved to the next hangar for work but there
is no gear to move it on. The planes are on jacks most of
the time they are in the hangar.

Steve signed off on the gear after he gave the gear
shop some inspections to do that he got from his buddy Lt
Smith on the next isle of the cube farm. He was a
mechanical systems engineer.

Date 0645 Saturday November 20, 1999
Rental House, village, OK

    Steve woke up to his cell phone ringing at an ungodly
hour on a Saturday morning. He was curious so he answered
it. It was Sgt McCay at the 507th Aerial Refueling Wing, the
Air Reserve Wing co-located on Tinker AFB. They also had
C/KC-135s and from time to time they would ask him to come
across the ramp to look at specific problems. Apparently
they were having a drill this weekend and they were trying
to generate up a flight but some airman had been screwing
around in the hangar and actually hit a golf ball with a 3
wood into the number one co-pilot window sending a spider
web across the expensive window. Of course none of that
story was shared with Lt Craken but MSgt McCay knew the Lt
was in charge of the 1418 wreck. MSgt McCay asked if he
could get the co-pilot number one window from 1418 and the
Lt said, "Yes go and get it", and then went promptly back
to sleep due to a late night with friends at Bricktown
downtown OKC.

Date 0705 Monday November 22, 1999
Rental house, Village, OK

Steve woke up in his rental house remembering a strange dream about 1418. The dream was something about a window being broken and someone getting a window for the wreck when suddenly the whole conversation from Saturday morning came back to him. ALL OF THE WINDOWS ON 1418 ARE CONDEMNED! Oh crap some aircraft is flying with a condemned window. *I'll call MSgt McCay from work*, he thought. *No I'll call him now he wisely reconsidered*. MSgt McCay answered the phone and Lt Craken said, MSgt I don't know what I was thinking. All the windows from 1418 are condemned. Where is that jet?" MSgt McCay said, "flying oversees I think. Let me check with the scheduler." It seemed like an eternity before he came back on the line. "It actually returned home early this morning." "Ok good", Said the Lt. "That jet is grounded until it can get another window". The MSgt said he would have to let his commander know and Lt Craken said, "I understand." Not a great start to the week.

Steve put on his Battle Dress Uniform (BDU) and drove to base. He looked at email and he had one from MSgt McCay. He really did not want to open it but he did.

It read:

Lt,

Base supply has a number one co-pilot window in stock
on base so we can remove and replace the condemned
window by tomorrow and not impact our schedule.

Thanks for telling us it was condemned.

MSgt McCay
Chief of maintenance
507 MXS

Steve religiously thanked his God for saving his bacon

and decided that no one else would try to get the windows.

He drove across base to the compass Rose and unlocked the

gate and drove up to 1418. He greeted the 654 CLSS crew

that were removing parts and asked them for a hammer. One

technician gave him a 2.5 pound brass hammer and a strange

look. Lt Craken climbed up to the flight deck and proceeded

to break every window. This action really freaked out the

maintenance crew and one said, "What the hell are you

doing?" So then Lt Craken had to tell them the story of the

condemned window. About half of the windows broke very

easily and the other half were structurally sound but there

was no way to tell which were good and which would shatter

in a pilot's face. That same technician said, "Never let

Officers have tools." Lt Craken laughed and said, "I have
to agree with you on that one!"

On his way out the crew entry door he noticed the two
data plates. The date plate tells what the aircraft is and
its serial number. One of the Technicians said, "You should
take that data plate and keep it so Steve asked him for a
screw driver. The Tech said, "Two tools in one day?" Lt
Craken pried the riveted data plates off and put them in
his pocket.

Date 1326 Thursday December 9, 1999
Cube Farm, Tinker AFB OK

Lt Craken was at his cube farm desk talking with Joe
the civilian petroleum engineer that sat next to him. They
were talking about building the Lt a computer for his
personal use from scratch when Margret, the section
secretary walked up and said, "You have a call from the
base fire chief on line one." Joe said, "Well that probably
isn't good." Steve said, "I agree, let's see what he
wants."

"Lt Craken this is Chief Wiggins and I'm going to need
a full report on the aircraft fire out on the compass rose.

I was told you are the on-scene commander of that wreck and all the work going on out there. I specifically need to know why and how two halon bottle fire extinguishers were used to put it out. I have to fill out two EPA reports." Lt Craken said, "I am providing engineer support for the 654 CLSS on what parts to remove but I will try to find out what is going on out there. What is your number?" The chief gave him his number and they hung up. Joe looked at the Lt and said, "Dang son, the base fire chief knows your name. I've worked at Tinker for 27 years and never got a call from the fire chief!" "Lucky me," Steve said sarcastically.

Steve went to his boss and explained the story because he needed some top cover for this amazing pile of crap he just stepped in. He also told him he had to go out to the compass rose where 1418 was parked.

Date 1326 Thursday December 9, 1999
Compass Rose, Tinker AFB OK

Lt Craken drove up to the compass rose and was relieved to see the CLSS technicians working because who knew their schedule. Sometimes he would go to the jet and it would be a ghost town but not today. Lt Craken walked up to the team lead and said, "hey I received a phone call

from the fire chief about a fire out here using two halon

fire bottles to put it out. What the hell?" TSgt Graves

said, "Oh right, about that, see we were trying out our new

plasma cutter instead of the K12 circular saw we usually

use and we happened to learn a couple of valuable lessons."

"And what were those?" asked the Lt. "Well number one the

plasma cutter is really just a lot of electricity used to

cut metal so if you cut across wires it will energize the

circuit. We thought that was really cool until we started

lowering the flaps and it hit SSgt Robinson in the head

almost knocking him out." "Holy Crap," said Steve. "Is he

OK?" "Umm…sure as much as he can be OK," said the TSgt.

"What was the second lesson?" the Lt asked showing his

annoyance. "Oh the second one, right that is the most

interesting phenomena", said the Sgt. Then the TSgt asked

the Lt a question. "Do you know what kind of metal the flap

track is made out of?" at first the Lt was like what does

that…then he quickly realized that some aircraft structures

were made out of magnesium, which is a strong light metal

but also a flammable metal that is very difficult to

extinguish. The Lt exclaimed, "YOU IGNITED MAGNESIUM?" "Yep

sure did and it took two bottles to put it out," said the

TSgt. "Did you think that might be something worth mentioning to me about so I would know it happened, you know so maybe if I got a call from the fire chief?", exclaimed the Lt. The TSgt said, "We had it out before the fire trucks got here." "Whoa the fire trucks rolled?" "Yes Sir" "OK, call me before the chief calls me, next time. Actually let's not do 'next time'," said the Lt. "Will do", said the TSgt.

Date 1025 Thursday February 10, 2000
Cube Farm, Tinker AFB OK

Lt Craken was zoning looking at his computer. He heard that Joe next to him was on the phone but wasn't listening. Joe hung up and turned to the Lt and said I have a contractor that wants the left wing from 1418. Steve said, "What? Oh just the left one? I have a two for one sale." Joe explained a contractor wanted the wing because the Air Force had hired them to figure out how to correct the "Top Coat" problem. The Top Coat was the final layer of paint that was inside the wing. Apparently it was peeling in huge pieces and normally that wouldn't be a problem except the C/KC-135 had integral wing fuel tanks which meant the paint peels were clogging up the fuel filters for the engines and could, in theory, cause a flame out. The wing section is wet with fuel with no rubber bladders. The body tanks all had rubber bladders. So the contractor NTP Specialties were asked to figure out a suitable replacement for the paint and they wanted a test bed. That meant the LT did not have to order the 654 CLSS to cut it up in small pieces to recycle. Unfortunately they did have to cut up the right wing.

This was great news. Another contractor ZAE Systems wanted the fuselage for training new engineers on aircraft structures.

All of the parts that were salvageable were off the air frame now so the only checklist item was to cut the wings off the fuselage and send the pieces to their new Homes.

Date 1300 Tuesday February 15, 2000
Compass Rose, Tinker AFB OK

Today is the last flight of C/KC-135 Tail Number 57-1418. Lt Craken led the five man team along with the two crane operators. He gave the order to cut off the wings. In the few weeks leading up to this event he talked at length

with some of the members of the 654 CLSS that were on the

previous team to cut up the last attrited C/KC-135 some

years earlier. TSgt Bosk said," I was inside the fuselage

holding the safety lanyard for TSgt Mosley who was cutting

the wing through the over-wing emergency exit hatch. He was

making the last cut on the wing and I heard a loud bang. I

pulled him back inside the cabin with all of my might and

saw the whole wing bucking up from the fuselage. It would

have killed him if I wasn't there." "let's not do that

again" said the Lt. We need to better place the cribbing to

equalize the weight of the wing so it does not move when it

is cut." So the team arranged the cribbing as the Lt called

for and proceeded to cut the wings off without incidence.

About thirty minutes later the flatbed semi-truck

arrived to take the fuselage off base to the contractor's

facility. The crane operators rigged the fuselage with two

slings and two cranes on either side. One crane lifted the

rear section of fuselage and the other lifted the front.

The cranes raised the fuselage about a foot off of its

cribbing. The enlisted guys started removing the railroad

ties that formed the cribbing like a huge Jenga game. They

were moving nonchalantly while working under the 50,000

pound fuselage "flying" overhead. "Let's move with a sense of urgency and get the cribbing moved!" exclaimed the Lt. Then Lt Craken jumped in and started moving the ties out of the way with the enlisted men. He firmly believed in leading by example. Ten minutes later the truck backed under the fuselage and the cranes lowered the wingless-tailless hulking carcass onto the flatbed trailer and C/KC-135 tail number 57-1418 landed for the last time in her 42 year career.

24.

Monday February 8, 2000
Hill AFB, UT

Lt Geronimo (Gerry) Ford stepped off the plane in terminal one at Salt Lake City Airport. Before today he had visited exactly three states that he did not live in; Texas where he went to Officer Candidate School, Oklahoma where he was currently stationed and Kansas because he had to drive through there to get home to Nebraska to visit. He was an officer in the Air Force yet before this Temporary Duty Yonder (TDY) he had never flown on an airplane. To say he was in awe of the mountains and this whole experience

was an understatement. He was super excited to be in Utah
and in the mountains.

He met up with Lt Craken and Lt Timmons at the car
rental counter and ended up with a green Honda Accord. They
were all drivers on the car so they rock paper scissors to
see who would drive. Gerry won or lost, depending on how
you look at it and so they loaded the trunk with luggage
and off to Hill AFB about 45 minutes' drive to the north.

They checked into billeting (military speak for hotel)
and received their room assignments. Steve went up to his
room and wasn't sure of what he would find. Some rooms were
really small and you shared a bathroom with the person in
the next room (like at Maxwell AFB) and they could lock you
out of the bathroom if they forgot to unlock your side. A
member could really luck out and get a two room suite with
private bathroom. His room was somewhere in between. He had
a bedroom with a private bathroom but there was a common
living room with another person. The other suite was
occupied and it was the instructor for the ABDR Engineer's
course…*Great sharing a suite with the teacher*, Steve
thought. *I guess I'll have to do the homework.*

Tuesday morning came and the three Lts drove to the
649 Combat Logistics Support Squadron (CLSS) and went to

the meeting room. It was packed with 12 engineer students

and the instructor Capt Evans, whom was Steve's suite mate

for two weeks.

There were engineers from Hill AFB, Robins AFB and the

three from Tinker AFB. Lt Craken was all jammed up at the

table and decided to roll his chair into the corner to get

more space to work. When he needed a desk he brought over

two large plastic tote containers and made a make shift

desk.

Tuesday through Friday the engineers started learning

their roles as ABDR engineers. Most of the engineers

learned the concepts to be used for battle damage in their

undergraduate schools but just needed to dust off the

equations. Two engineers were totally lost by day two. Capt

Evans wasn't sure why they were having so many problems

with the introductory material they should have already

known from their degrees but he tried to help them as much

as possible.

The curriculum started off with the ABDR repair

methodology. The overall steps for repair design for the

engineers are:

1. Evaluate the Damage.

2. Gather Material Properties.

3. Calculate Original Structural Capabilities

4. Choose Repair Components and Verify Structural Soundness.

5. Choose Fastener Type, Size and Location.

6. Draw Repair Diagrams

7. Sanity Check

8. Write Repair Instructions

9. Archive Repair Strategy.

One of the material properties that the engineer needs is area properties. The engineer needs the area centroid for the cross section of the damaged piece of structure and the area moment of inertia. The centroid is where you could balance the weight of the piece of structure at one point. The moment of inertia is basically the resistance to rotation. It is easier to spin around object with the mass mostly at the center than it is to spin a square or other polygon with the mass at the outside corner.

The reason the engineer needs this information about the damaged section is to ensure the repair has enough strength to carry an unknown bending moment. See figure 10 for some sample cross section centroids for typical aircraft structure.

Aircraft structure can be loaded in the following ways; bending, compression, tension, torque and shear. For example wing skin is designed to withstand the bending loads in flight and on landing. In flight the upper wing skin is in compression and the bottom wing skin is in tension due to the aerodynamic loads pulling the wing up and anything attached to the wing like the fuselage. The wing structure is in bending (up or down depending on where the structure is above and below neutral axis and mission profile - taxiing or flying). The neutral axis is where the structure has zero bending.

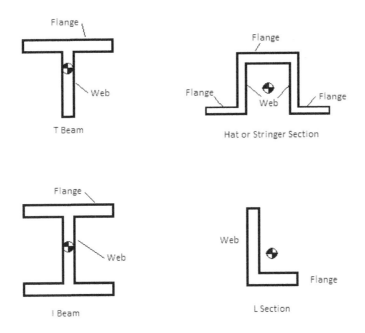

Figure 9 Common cross sections of aircraft structures

The exact opposite is true when the aircraft is on the ground. The upper wing skin is in tension and the lower wing skin is in compression and the structure flexes a lot.

By Friday everyone, including the instructor, was ready for the weekend. Capt Evans asked the class if anyone wanted to go skiing at the local resorts. A large majority did want to go so they planned to go all day Saturday at The Solitude Resort.

Steve and Gerry decided they needed some gear for skiing so they went to a local mega sports store and checked out the gear they needed. They both needed hats, goggles and gloves. Steve found a bargain bin of bright yellow gloves for five dollars and he picked up a pair. Next were hats. Steve found some and went for a regular ski cap but Gerry found a reproduction WWII pleather bomber hat and put it on as a joke. Steve somehow convinced him that he looked really cool wearing the bomber hat that had the faux fur ear flaps.

The class met in the parking lot of Solitude Ski Resort and Steve decided he should take a lesson because he had never been on skis before. Capt Evans said, "No you don't need a lesson. I will teach you." Steve listened to his instructor and opted out of the lesson… that was a big mistake. Steve fell no less than three times trying to go up the tow rope for the bunny hill. Kids as young as three years old had mastered the tow rope but not Steve or Gerry.

Once they were on the "top" of the bunny hill Capt Evans tried to give Steve and Gerry a lesson. He was not a good ski instructor. He had Steve and Gerry snow plowing but did not tell them how to stop or turn. So Capt Evans started going down the hill trying to show the duo how to ski and he said you try. So Steve went for it and ended up right behind the Captain, Steve's chest to Capt Evans back with Steve's skis inside Capt Evans. It looked very weird. Steve couldn't stop so he just sat down. Steve thought this guy is an idiot. I'll have to teach myself and he did. He picked it up fairly quickly and wasn't afraid of dying or running up inside someone's skis and backside making a 'new friend'.

They skied for an hour and in the process they lost Gerry. Steve, Sarah and the Capt were on the Moon Beam Express going up to some green runs and they were asking each other if they had seen Gerry. Right after someone asked where he was, Steve saw some bright yellow gloves and that stupid looking bomber hat. He was flying down the Little Dollie run close to the lift and Steve said,

"There's Gerry". Capt Evans Yelled, **"Hey, Gerry, meet us at the lodge!"** Hearing his name Gerry looked up and proceeded to face plant and become the mother of all garage sales loosing that goofy hat down the steep slope. Everyone on the lift was dying laughing.

The next week they solidified all the things they learned in the previous week and started designing repairs. There was an old F-16 right outside the conference room that they used to make BDR repairs on. Hill AFB is the depot for fighter aircraft such as the F-15 Eagle, A-10 Warthog and the F-16 Viper.

During the class they had regular checks to determine they understood the material. See the example problems.

Steve and everyone else passed the course except the two engineers that were basically lost the whole time.

1) For the J-Section shown, calculate the following (ignore the fillets in your calculations)
   a) Centroid
   b) Moment of inertia about the centroidal x-axis, $(I_{xx})_c$
   c) Moment of inertia about the centroidal y-axis, $(I_{yy})_c$
   d) Centroidal product of inertia, $(I_{xy})_c$
   e) Principle moments of inertia and directions
   f) Radius of gyration, $(\rho_x)_c$
   g) Radius of gyration, $(\rho_y)_c$

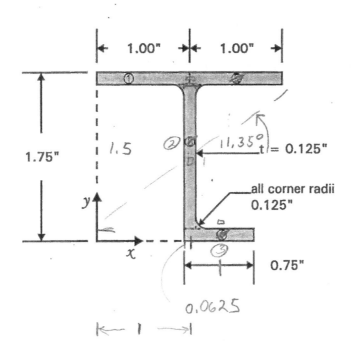

INTRODUCTION TO ABDR ENGINEERING
PROGRESS CHECK #3 (SOLUTIONS)

SECTION	x_i	y_i	A_i	A_ix_i	A_iy_i	$I_{xx,i}$	$I_{yy,i}$	$A_ix_i^2$	$A_iy_i^2$	$A_ix_iy_i$
1	0.500	1.688	0.125	0.063	0.211	0.000	0.010	0.031	0.356	0.105
2	1.500	1.688	0.125	0.188	0.211	0.000	0.010	0.281	0.356	0.316
3	1.000	0.875	0.188	0.188	0.164	0.035	0.000	0.188	0.144	0.164
4	1.313	0.063	0.094	0.123	0.006	0.000	0.004	0.161	0.000	0.008
TOTALS			0.531	0.561	0.592	0.036	0.025	0.661	0.856	0.594

a) xbar = 1.055
   ybar = 1.114

b) $(I_{xx})_c$ = 0.232

c) $(I_{yy})_c$ = 0.096

d) $(I_{xy})_c$ = -0.031

e) $(I_{xx})_p$ = 0.239
   $(I_{yy})_p$ = 0.089
   θ_p = 12.1

f) $(\rho_x)_c$ = 0.661

g) $(\rho_y)_c$ = 0.424

$$I_{P1,P2} = \frac{I_{xx}+I_{yy}}{2} \pm \sqrt{\left(\frac{I_{xx}-I_{yy}}{2}\right)^2 + I_{xy}^2}$$

$$\tan(2\theta) = \frac{-2I_{xy}}{I_{xx}-I_{yy}}$$

Equations to find the principle moments of inertia

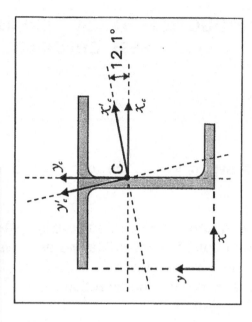

Element	A_i	$\bar{x}_i$	$A_i\bar{x}_i$
1	0.25	1	0.25
2	0.1875	1	0.1875
3	0.094	1.3125	0.1234
	0.5315		.5609

$$\bar{x} = \frac{\Sigma A_i\bar{x}_i}{\Sigma A_i} \Rightarrow \boxed{\bar{x} = 1.055 \text{ in}}$$

Element	A_i	$\bar{y}_i$	$A_i\bar{y}_i$
1	0.23	1.69	0.423
2	.1875	0.875	0.164
3	0.094	0.0625	0.0059
	.5315		0.593

$$\bar{y} = \frac{\Sigma A_i\bar{y}_i}{\Sigma A_i} = \boxed{\bar{y} = 1.116 \text{ in}}$$

0833

Ele	A_i	$\bar{y}_i$	b_i	h_i	I_{xx_i}	$A_i\bar{y}_i$	$A_i\bar{y}_i^2$
1	.25	1.69	2	.125	0	.423	.715
2	.1875	.875	.125	1.5	0.035	.164	.144
3	.094	.0625	.75	.125	0	.0059	0
	.532				.035	.593	.859

$$I_{xx_c} = \Sigma(I_{xx_i}) + \Sigma A_i\bar{y}_i^2 - \bar{y}\left(\Sigma A_i\bar{y}_i\right)$$

$$\underset{\wedge}{.035} \quad + \quad .859 \quad - 1.116\,(.593)$$

$$\boxed{I_{xx_c} = 0.2322 \text{ in}^4}$$

Ele	A_i	X_i	b_i	h_i	I_{yyi}	A_iX_i	$A_iX_i^2$
1		1	2	.125	.0833	.25	.25
2		1	.125	1.5	0	.1875	.1875
3			1.313	.75	.125	.0044	.1234	.162
	.5315				.0877	.5609	.5995

$.0833b^3h$

$$I_{yy_c} = \Sigma(I_{yyi}) + \Sigma A_i \bar{X}_i^2 - \bar{X}(\Sigma A_i \bar{X}_i)$$

$$.0877 + 0.5995 - 1.055(.5609)$$

$$\boxed{I_{yy_c} = 0.0955 \text{ in}^4}$$

$A_i \bar{X}_i \bar{Y}_i \qquad \bar{X}\bar{Y} = 1.177$

.4225
.164
.01

0.597

$$I_{xy_c} = \Sigma(\cancel{I_{x'y'i}}^0) + \Sigma A_i \bar{X}_i \bar{Y}_i - \bar{X}\bar{Y}(A_i)$$

rectangular elements

$$0.597 - 1.177(.5315)$$

$$\boxed{I_{xy_c} = -0.0286 \text{ in}^4}$$

$$I_{P_1, P_2} = \underbrace{\frac{I_{xx_c} + I_{yy_c}}{2}} \pm \sqrt{\left(\frac{I_{xx_c} - I_{yy_c}}{2}\right)^2 + I_{xy_c}^2}$$

$$.1639 \pm 0.074$$

$$I_{P_1} = 0.238 \text{ in}^4$$

$$I_{P_2} = 0.0899 \text{ in}^4$$

$$\theta_p = \frac{1}{2} Arc\tan\left(-\frac{2 I_{xy_c}}{I_{xx} - I_{yy}}\right)$$

.4184

$$\boxed{\theta_p = 11.35°}$$

$$(\rho_{xx_c}) = \sqrt{\frac{I_{xx_c}}{A}} = 0.661 \frac{in^4}{in^2} \, in^2$$

$$(\rho_y)_c = \sqrt{\frac{I_{yy}}{A}} = 0.424 \, in^2$$

INTRODUCTION TO ABDR ENGINEERING
PROGRESS CHECK #5

NOTE: Employ a 15% fitting factor in all margin of safety calculations.

1. For the given diagram (p = 725 lbs.),
   a. calculate the shear and bearing allowables
   b. determine whether the joint can sustain the design load and calculate the margins of safety

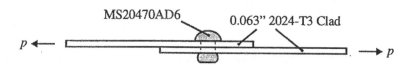

2. For the given diagram (p = 850 lbs.),
   a. determine the joint allowables for both the fastener head and tail
   b. determine whether the joint can sustain the design load and calculate the margins of safety

$F_{su} = 125\ 000$
$F_{bry} = 84\ 000$

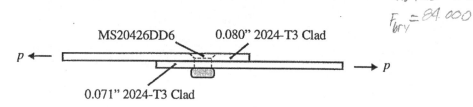

3. For the given diagram (p = 8,000 lbs.),
   a. calculate the shear and bearing allowables
   b. determine whether the joint can sustain the design load and calculate the margins of safety

$F_{bru} = 146\ 000$
$F_{bry} = 110\ 000$

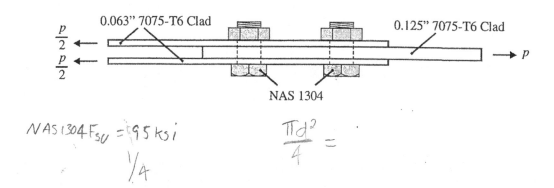

$NAS\ 1304\ F_{su} = 95\ ksi$

$\frac{1}{4}$

$\frac{\pi d^2}{4} =$

4. For a one inch wide strip (into and out of the page) of the stacked patch shown,
   a. calculate the ultimate joint strength (ignore fitting factor at this stage)
   b. assuming a design load of 7800 lbs., calculate the margin of safety

$F_{tu} = 146000$ .125
$F_{bry} = 110000$

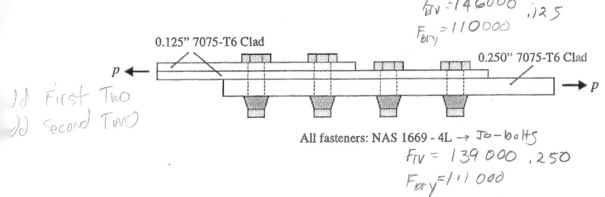

0.125" 7075-T6 Clad

0.250" 7075-T6 Clad

$p \leftarrow$

$\rightarrow p$

a) First Two
b) Second Two

All fasteners: NAS 1669 - 4L → Jo-bolts
$F_{tv} = 139\,000$ .250
$F_{bry} = 111\,000$

5. A technician is installing a 0.063" Clad 2024-T3 patch on the fuselage of an aircraft whose factory skin is 0.050" Clad 2024-T3. The fastener he is using is the CR3243-6. He somehow gets confused and makes all the edge/end distances on outside row of fasteners equal to 3/16" (i.e. e/D = 1.0). Calculate the fastener allowable for the outside rows.

$F_{bru} = 121\,000$
$F_{bry} = 82\,000$ .05

$F_{bru} = 125\,000$
$F_{bry} = 84\,000$ .063

$\frac{p}{2} \downarrow$ $\downarrow \frac{p}{2}$

e L

D

0.063"
2024-T3 Clad

p

$P_{Allow} = 2 F_{su} \cdot t (e - .383 D)$
           in
          sheet

6. Find the maximum shear flow this fastener row can sustain

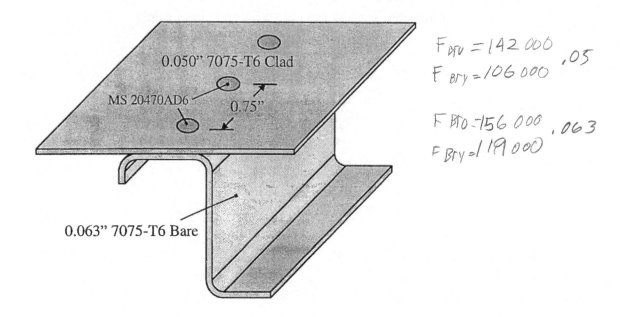

0.050" 7075-T6 Clad

MS 20470AD6

0.75"

0.063" 7075-T6 Bare

$F_{bru} = 142\,000$
$F_{bry} = 106\,000$ .05

$F_{bro} = 156\,000$ .063
$F_{bry} = 119\,000$

For the lug mounting plate shown
a. Determine the loads in all four fasteners
b. Which fastener is the most highly loaded?
c. Choose a fastener for the mounting plate. Explain your choice

Steel

$F_{bru} = 331$ ksi
$F_{bry} = 246$ ksi

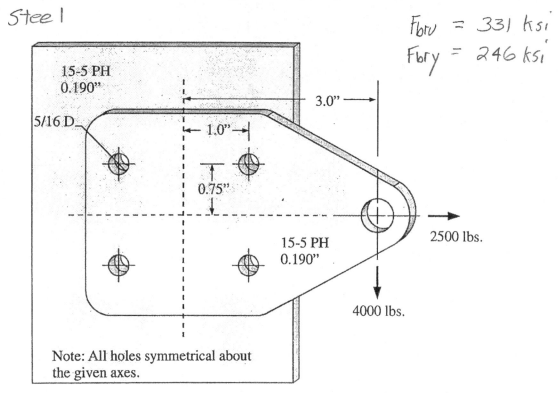

15-5 PH
0.190"

5/16 D

3.0"

1.0"

0.75"

15-5 PH
0.190"

2500 lbs.

4000 lbs.

Note: All holes symmetrical about the given axes.

1) For a protruding head rivet, we must find the shear strength of the rivet and the bearing strength of the sheet in order to determine whether or not this joint can safely sustain the design load.

a) The uncorrected ultimate shear strength is:

$$F_{su} A_s = 862 \text{ lbs}$$

The shear allowable is calculated using the equation

$$P_{su} = nc(F_{su}A_s)$$
$$= (1)(1.0)(862 \text{ lbs}) = 862 \text{ lbs}$$

For the bearing allowable, we find

$$P_{bru} = F_{bru} Dt$$
$$= (125 \text{ ksi})(0.191 \text{ in})(0.063 \text{ in}) = 1,504 \text{ lbs}$$

$\uparrow$ MIL-HDBK-5G

b) Now, we calculate the margins of safety

Shear: $MS = \dfrac{862 \text{ lbs}}{1.15(725 \text{ lbs})} - 1 = \underline{\underline{3\%}} \Leftarrow$

Bearing: $MS = \dfrac{1,504 \text{ lbs}}{1.15(725 \text{ lbs})} - 1 = \underline{\underline{80\%}} \Leftarrow$

2) This joint uses a machine countersunk DD rivet and joins two sheets of unequal thickness. Therefore, we must evaluate the head and the tail independently.

a) Head

MS20426DD6 in Clad 2024-T3, 0.080" : 992 lbs

Tail (evaluate as protruding head solid)

$$F_{su} A_s = 1175 \text{ lbs} \quad (DD6 \text{ rivet})$$

$$P_{su} = nc(F_{su}A_s) = (1)(1.0)(1175 \text{ lbs}) = 1175 \text{ lbs}$$

2) (cont.)

   Tail (cont.)

$$P_{bru} = F_{bru} D t = (125 \text{ ksi})(0.191 \text{ in})(0.071 \text{ in})$$
$$= 1,695 \text{ lbs.}$$

b) The only margin of safety we really need is that for the head. However, for the sake of completeness, I will calculate margins of safety for both the head and tail.

Head
$$MS = \frac{992 \text{ lbs}}{(1.15)(850 \text{ lbs})} - 1 = \underline{1\%} \impliedby$$

Tail

Shear: $MS = \dfrac{1175 \text{ lbs}}{(1.15)(850 \text{ lbs})} - 1 = \underline{20\%} \impliedby$

Bearing: $MS = \dfrac{1695 \text{ lbs}}{(1.15)(850 \text{ lbs})} - 1 = \underline{73\%} \impliedby$

3. In this problem, we must take into account the fact that the joint is in double-shear and that there are two fasteners. First, we will determine the shear and bearing allowables.

a) From the 1-1A-8, an NAS1304 has an ultimate shear strength of   — Double shear
$$P_{su} = (4,650 \text{ lbs.})(2) = 9,300 \text{ lbs}$$

For clad 7075-T6 0.063" – 0.187", $F_{bru} = 146$ ksi (A). The middle sheet is therefore more critical in bearing than the outer sheets (combined). 0.125" < 2(0.063")

$$P_{bru} = (146 \text{ ksi})(0.250 \text{ in})(0.125 \text{ in}) = 4562 \text{ lbs}$$

3. (cont.)

b) From the results in part a), we can see the fasteners (bolts) are bearing critical. For a concentrically loaded joint with fasteners of the same type and size, the load on each bolt is found by dividing the total load on the joint by the number of bolts.

$$P_f = \frac{P}{n} = \frac{8000\,lbs}{2} = \underline{4,000\ lbs}$$

Now, calculate the margins of safety.

Shear: $MS = \dfrac{9,300\,lbs}{(4,000\,lbs)(1.15)} - 1 = \underline{\underline{102\%}} \Longleftarrow$

Bearing $MS = \dfrac{4562\,lbs}{(4,000\,lbs)(1.15)} - 1 = \underline{\underline{-1\%}} \Longleftarrow$

4. A joint is only as strong as the sum of the allowables of the fasteners in the joint. Hence, we must find two allowables: one for the jo-bolts through all layers, and one for the jo-bolts only passing through the fatigue doubler only.

a) First, the jo-bolts through all sheets

NAS1669 in 0.250" 7075-T6 CLAD = 4,500 (shear limited).
NAS1669 in 2 x 0.125" 7075-T6 CLAD :
                              monolithic: 4,500 (shear limited)

Now, the allowable for the jo-bolts in the fatigue doubler.

NAS1669 in 0.125" 7075-T6 CLAD = 3,820 lbs (bearing)

$$P_{allow} = 2(4,500) + 2(3,820) = \underline{16,640\ lbs.}$$

b) If we assume all the jo-bolts are there to carry load,

$$MS = \frac{16,640\,lbs}{(7800\,lbs)(1.15)} - 1 = 86\%$$

4.b) (cont.)

In most cases, however, we only consider the fasteners passing through all sheets to be load-bearing in order to improve fatigue characteristics of the patch. Thus,

$$MS = \frac{9,000\ lbs}{(1.15)(7800\ lbs)} - 1 = \underline{\underline{0\%}} \Leftarrow$$

5. With an edge distance as small as 1D, The sheet metal will not be able to develop full bearing strength in the sheet or maximum shear strength in the fastener. Therefore, we will calculate the load at which tear-out failure will occur.

From eqn. 6-13 in the new engineering handbook,

$$P = 2F_{su}t\ (e-0.383D)$$

$$e = 0.207"\qquad D = 0.207"\qquad F_{su} = 39\ ksi\qquad t = 0.063"$$

$$P_{allow} = 2(39\ ksi)(0.063\ in)\left[0.207 - (.383)(0.207)\right]$$

$$= \underline{\underline{628\ lbs}}$$

6. From eqn. 6-24 in the new engineering handbook

$$q = \frac{(P_{allow})(\#\ Rows)}{P}$$

In this problem, we know the fastener pitch is 0.75", and we have a single row of fasteners.

$$P = 0.75"\qquad \#\ Rows = \underline{1}$$

Therefore, we need only find the maximum allowable load each rivet can carry to reverse engineer the shear flow in this attachment line.

MS20470AD6 is a 3/16" diameter AD protruding, tension head rivet.

6. (cont.)

From Table C.2.2 in the new engineering handbook,

$$F_{su}A_s = 862 \text{ lbs}$$

From Table C.2.4, we find the correction factor (single-shear) for a protruding head AD6 rivet in 0.050" sheet:

$$c = 0.970$$

Next, we plug into Eqn. 6-17.

$$P_{su} = n\, c\, F_{su}A_s = (1)(0.97)(862 \text{ lbs}) = \underline{836 \text{ lbs}}$$

We now have the shear strength, but we must also find the bearing allowables for the stringer and the sheet.

Stringer

$$0.063"$$

Eqn. 6-5 : $P_{bru} = F_{bru}\, D t$ $\Rightarrow F_{bru} = 156 \text{ ksi}$

$$P_{bru} = (156 \text{ ksi})(0.191 \text{ in})(0.063 \text{ in.})$$ $F_{bry} = 119 \text{ ksi}$

$$= \underline{1877 \text{ lbs}}$$ $1.5\, F_{bry} = 178 \text{ ksi}$

Sheet

$$0.050"$$

$$P_{bru} = (142 \text{ ksi})(0.191 \text{ in})(0.050 \text{ in.})$$ $\Rightarrow F_{bru} = 142 \text{ ksi}$

$$= \underline{1356 \text{ lbs}}$$ $F_{bry} = 106 \text{ ksi}$

$$1.5\, F_{bry} = 159 \text{ ksi}$$

The shear strength of the fastener is thus the weak link.

$$P_{allow} = 836 \text{ lbs,}$$

$$q = \frac{(836 \text{ lbs})(1)}{\underbrace{0.75"}_{\text{Fastener pitch}}} = \underline{\underline{1,115 \frac{\text{lbs}}{\text{in}}}}$$

7. From section 6.6 in the engineering handbook, The steps to take to solve This problem are

a) a) Find the center of resistance
   b) Determine the reactions at this centroid : P, V, M
   c) Calculate shear forces due to P & V
   d) Calculate shear forces due to M
   e) Find x and y components of the shearing forces
   f) Find resultant shearing forces on fasteners

Step 1 → Done

Step 2

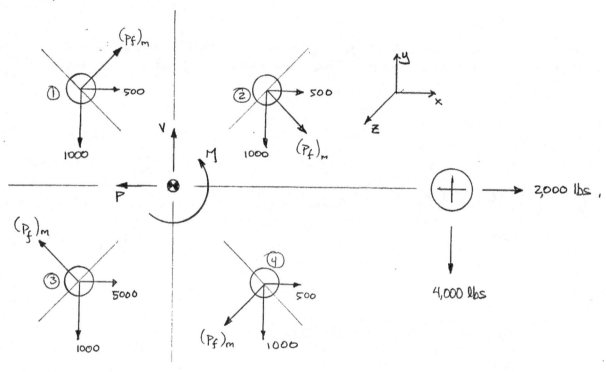

$$\Sigma F_x = -P + 2000 \text{ lbs} = 0 \qquad P = 2,000 \text{ lbs}$$

$$\Sigma F_y = V - 4000 \text{ lbs} = 0 \qquad V = 4,000 \text{ lbs.}$$

$$\Sigma M_{cg} = M - (4000 \text{ lbs})(3 \text{ in.}) \qquad M = 12,000 \text{ in.-lbs.}$$

Step 3

We will assume all fasteners will be of the same size and type.

$$(P_f)_P = \frac{P}{n} = \frac{2000 \text{ lbs}}{4} = 500 \text{ lbs.}$$

7. (cont.)

Step 3 (cont.)

$$(P_f)_v = \frac{V}{n} = \frac{4000 \text{ lbs}}{4} = 1,000 \text{ lbs}.$$

Step 4

For this portion of the problem, we will set up a table.

FASTENER	$\bar{x}_c$	$\bar{y}_c$	r	r^2	$[(P_f)_m]_x$	$[(P_f)_m]_y$
1	1.0	0.75	1.25	1.562	1440	1920
2	1.0	0.75	1.25	1.562	1440	1920
3	1.0	0.75	1.25	1.562	1440	1920
4	1.0	0.75	1.25	1.562	1440	1920

TOTAL   6.25

$$[(P_f)_m]_{i,x} = \frac{M}{\sum_i r_i^2}(\bar{y}_c)_i \qquad [(P_f)_m]_{i,y} = \frac{M}{\sum_i r_i^2}(\bar{x}_c)_i$$

Step 5 & Step 6

For this portion, we will also construct a table. Additionally, we must remember to adjust the signs of the results from steps 3 & 4 to account for their actual direction.

FASTENER	$(P_f)_v$	$(P_f)_P$	$[(P_f)_m]_x$	$[(P_f)_m]_y$	ΣF_x	ΣF_y	P_f
1	-1000	500	1440	1920	1940	920	2147
2	-1000	500	1440	-1920	1940	-2920	3506
3	-1000	500	-1440	1920	-940	920	1315
4	-1000	500	-1440	-1920	-940	-2920	3068

$$\Sigma F_x = (P_f)_P + [(P_f)_m]_x \qquad \Sigma F_y = (P_f)_v + [(P_f)_m]_y$$

$$P_f = \left[(\Sigma F_x)^2 + (\Sigma F_y)^2\right]^{1/2}$$

b) Fastener ② is the most highly loaded.

7. (cont.)

c) Based on my personal knowledge of fasteners, which is admittedly somewhat limited, I will choose the NAS1305 bolt.

To prove it is sufficient, I will calculate the shear strength of the fastener, the bearing strength of the steel plates, and appropriate margins of safety.

For a 95 ksi steel bolt w/ $D = 5/16''$

$$F_{su} A_s = (95 \text{ ksi}) \frac{\pi \left(\frac{5}{16}\right)^2}{4} = 7286$$

$$P_{su} = n \not\ell^{\frac{1}{2}} F_{su} A_s = 7286 \text{ lbs}$$

$$MS = \frac{P_{su}}{k \, P_f} - 1 = \frac{7286}{(1.15)(3506)} - 1 = \underline{\underline{81\%}} \Leftarrow$$

The 0.190" steel plate will provide a considerable bearing strength.

MIL-HDBK-5G : AMS 5862 (15-5 PH) plate Tbl. 2.6.6.0(c)

$F_{bru} = 331$ ksi $F_{bry} = 246$ ksi   (Use F_{bru})

$$P_{bru} = F_{bru} D t = (331 \text{ ksi})\left(\frac{5}{16} \text{ in}\right)(0.190 \text{ in})$$

$$= \underline{19,653 \text{ lbs}}$$

$$MS = \frac{P_{bru}}{(1.15) P_f} - 1 = \frac{19653}{(1.15)(3506)} - 1 = \underline{\underline{\text{AMPLE}}} \Leftarrow$$

Since I have proven this fastener will work for the most highly loaded fastener, the other fasteners are covered by simple logic!

25.

Date 0930 Wednesday March 10, 2000
OC/ALC Tinker AFB, OK

Lt Craken was at his desk feigning work because he had

a heavy week before this one with computer based training

and several field calls that he had to design repairs for.

Lucy Sheen, who was a civilian, came up to the Lt's desk

and started asking how he was. Lucy was a mother figure to

all the new Lts in the SPO and quite feisty. She was an

Item Manager for the C/KC-135. An Item manager keeps track

of a set of parts, or items that go on the aircraft. There

are thousands of parts that make up the whole aircraft.

They talked a little while about the weekend and then got down to business. The C/KC-135 had airframes that were well over 40 years old and the logistics to keep them flying costs millions of dollars a year. So it was no surprise that there was an airframe stationed at McConnell AFB Kansas that was grounded because it needed a main landing gear left hydraulic pump that hasn't been manufactured for 20 years with zero in supply. All the while the LT was wondering what in the heck this had to do with him because he was structures not systems, so he said as much. Lucy said she knew that but there was indeed one each left hydraulic pump installed on the C/KC-135 static museum piece on concrete blocks at the Tinker Main gate and all the paper work had been filled out but they needed someone to pull it off. The Lt thought for a minute and then said, "If you buy a couple cases of beer I can get some technicians over at the CLSS to go and remove it. Then I can take it to supply for shipment to McConnell."

So Lt Craken went to the Class Six (Liquor store on base) and picked up a couple of cheap cases of beer. Then he drove to the East Annex, Home of the 654 Combat Logistics Support Squadron (CLSS). He walked into the entry way and there were a couple of NCOs hanging out and saw the

beer. One NCO said," looks like the Lt needs some manual
labor done", and she was right. Lt Craken went to the
C/KC-135 section and found a couple of structures
technicians and told them what he needed. They said,"
we'll take care of it", and they did. By early morning the
next day the pump was leaking Mil-H-5606 red hydraulic
fluid all over the Lt's desk and he didn't care.

Lt Craken transported the pump to massive logistics
center in the middle of the base and dropped it off at the
desk. This building had to be the government warehouse that
the Ark of the Covenant was placed in at the end of the
Indiana Jones Movie. The logistician took care of it from
there. Mission Complete!

26.

Date 1500 Wednesday March 10, 2000
763 Aircraft Maintenance Unit Offutt AFB, Nebraska

    MSgt Calc, The maintenance Supervisor, walked into

Hangar number two where C/KC-135 Tail number 56-2451 was

undergoing its annual phase inspection. There was severe

corrosion in the number seven fuselage longitudinal

stringer. The aircraft was jacked up off the ground to

facilitate maintenance. If there is a major fuselage repair

the aircraft has to be jacked so when the damaged area was

removed or cut away the other load paths don't fail.

Structural failure could look like bending, crippling,

buckling, broken fasteners, corrosion, elongated fastener holes and hole shear-out where fasteners were pulled from the joining structure to name a few.

TSgt Brady and Airman Rye were looking at the stringer and had just finished cutting away the corrosion. Aircraft built after World War II were mostly made of aluminum with some parts made of steel, magnesium and titanium. Aluminum corrosion or aluminum oxide (Al_2O_3) is a white powder and it has three times the volume of aluminum in its solid metal form. Since the corrosion has more volume it can crack the metal but makes it easy to identify.

MSgt Calc asked the two technicians how the repair was going. TSgt Brady said," The repair is in the book but the limits are outside the repair so we will need to go back to the OC/ALC for repair instructions. Another problem is I checked bench stock and we are out of this hat section stringer." MSgt Calc said, "Well that's a real big freaking problem. This bird is due to the sandbox early next week." He was talking about next week's mission to the Middle East. TSgt Brady said, "Supply is way too slow for that." MSgt Calc said, "I wish we could fly our other bird to Tinker and get what we need. That is a two hour flight." TSgt Brady said," There is a sharp LT in the OC/ALC that could possibly get us what we need." MSgt Calc said, "What?

A sharp LT? They make those?"

Date 1515 Wednesday March 10, 2000
OC/ALC Tinker AFB, Ok

Lt Craken answered the phone, "135 SPO this is Lt

Craken." "Lt this is TSgt Brady at Offutt", "Hi Sargent.

How are you?" "Pretty good", said TSgt Brady. He told the

LT what he needed and said he would email the part numbers

to the two different hat stringers they needed for the

repair in the technical order. The Lt said he would call

the TSgt tomorrow when he had the parts. The Lt was fairly

sure that the parts were down on the depot line and he

could just go pick them up in the morning. TSgt Brady also

said, "The stringer repair is in the books but the

allowable damage is outside tech order guidance. Is there

any way you could design a repair for us in a day?" The Lt

said, "No problem, send me dimensions and a pic if

possible." The TSgt said he would have them in the hour.

Date 0745 Thursday March 11, 2000
OC/ALC Depot Line Tinker AFB, Ok

Lt Craken walked down to the depot floor shortly after coming into the cube farm that morning. He walked into bay three where tail number 64-17241 was getting its every-five year depot rebuild. The C/KC-135 would almost be a brand new bird when it left Tinker in 7 months. He then went over to the structures bench stock and went shopping. He needed a BAC 1498 Channel hat section in an 8 foot length and a BAC 1497 channel hat section also in an 8 foot length. The BAC 1497 nests inside the BAC 1498 to build up the repair. See Figure 11 for repair and sketch. Both hat sections are made from 2024-T3 Aluminum. The 2024 is the recipe or type of aircraft grade aluminum. The "T-3" is the heat treat recipe so the stringer is strong but also is relatively ductile. If the stringer was brittle it would break instead of flex. Aircraft structures are made to flex.

The Lt remembered when he was in college at MSU his class took a field trip to an airplane bone yard in Greenville, MS where there were retired Boeing 747s. The class was able to go to the wing tip and jump on the wing. The wing flexed with the weight of one person a few feet!

Lt Craken found the parts he needed and began walking back up to his cubicle. No one said a word but he did get some strange looks. He called the TSgt at Offutt and told

him he had the parts. Now to design the repair.

Date 1000 Thursday March 11, 2000
Offutt AFB, NE

TSgt Brady let operations (Ops) know the parts were

ready two hours ago and the ground crew had worked hard to

get the C/KC-135 ready for take-off. The flight crew

arrived and Ops instructed them that a sharp Lt would meet

them on the taxi-way at Tinker AFB and deliver parts for

the repair of the phase bird. The aircraft commander,

Captain Petty said, "A sharp LT…Do they make those?" Ops

said, "Apparently there is one…at Tinker."

Tail number 58-4351 took off of runway 12 at 1025 for the

two hour flight to Tinker AFB.

Date 1225 Thursday March 11, 2000
Tinker AFB, OK Taxi way Foxtrot

Lt Craken loaded up in a blue Base Ops pickup with the

eight foot stringers in the back. The Lt put in his

hearing protection because they were going to meet the jet

with all four engines at idle and this C/KC-135 had the

engine upgrade to the large and efficient General Electric

CFM-56 engines.

The jet stopped on the taxi way and the truck drove to

75 feet in front of the nose. The boom operator climbed

down the crew ladder to meet the LT. C/KC-135 aircraft are

in-flight refueling aircraft and they use a large boom on

the tail of the aircraft to deliver fuel to other aircraft.

The boom operator "flies the boom" and performs other

inflight duties.

The Lt brought up the stringers and was not sure they

would go through the crew entry door beneath the flight

deck. The C/KC-135 also has a large cargo door several

feet aft of the flight deck. After some finagling and

Tetris, the crew got the stringers up the crew entry and

stowed them in the cargo area. The boom operator asked the

LT to climb into the flight deck and he did so. TSgt Brady

met the LT and gave him a whole bunch of corn husker swag,

including the unit patch and hat. Lt Craken said thanks and

the aircraft commander thanked the LT for the parts and

said, "Good job being a sharp LT." The joke was lost on

the Lt but he said," No problem", and climbed back down the

ladder and loaded back up in the truck. Then he watched the

135 roll to the runway, set brakes, run-up the large

engines, release brakes and depart to the south, turn and head back north to the land of the Huskers. He would head back to the office and email the repair disposition to the Sargent that just left. Another mission complete.

Repair Materials
  BAC 1497 HAT
  BAC 1498 HAT

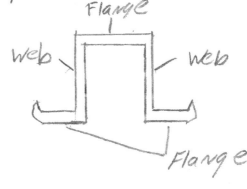

web / Flange \ web

Flange

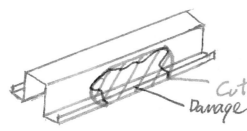

Cut Damage Area out.

Damage

Cut a Section of BAC 1498 to replace cut out area

Nest A BAC 1497 inside original structure

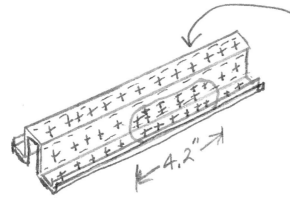

Blind Side Same pattern
↕ 1.25"

t = 0.090"

←4.2"→

Damage Area = 5.25 in²   Repair Area >> 5.25 in²

Fasteners Hi-Lok   0.25 Diameter in t=0.090

  2430 lba   NAS 1447

Effective Fasteners 24 = 58,320 lba
Good for Fatigue.

27.

Date 0930 Monday March 15, 2000
C/KC-135 SPO cube farm, Lt Craken's desk

    Lt Jones bounced up to Steve's cube along with Lt Rob
Smelt and was in a crazy good mood for a random Monday.
Steve said with a head nod, "what's up my hommies?" Lt
Jones exclaimed, "I have an inside contact over at the
Oklahoma City Air National Guard (ANG) unit that said they
are flying to St. Croix April 3 and they have a sign up for
space-A!" Rob said, I don't know what "Space-A" is but it
sounds cool". Lt Craken explained, "Space-A is a military
flight that you can fly on for free if they have room".

Rob was a strange dude for sure. At first impression he seemed to be a normal officer but Steve had gotten to know him a little over the past six months. One thing that was really cool about him was that he owned his own small airplane, an old bonanza and they had flown a cross country to get the proverbial $100 hamburger. Another factoid is he was really into pornography. That was not cool to Steve because he was a far right wing Christian and left that stuff alone. Rob even looked into becoming a porn star and that seemed very weird behavior for an officer in the United States Air Force. Other than that he was a fairly cool dude and Steve hung out with him from time to time.

Steve looked at Tiffany and asked, "What do they fly over there at the ANG base? C-130 cargo planes?" she said, "Yes and we are going!" "I'll have to look at my leave and probably have to take advanced leave." "Me too" the other officers agreed. "Sweet, let's do it, said Steve.

The two other officers left to make their rounds. Steve got the luck of the draw with job assignments and actually had work he was expected to do. Typically that is not the case with new 2nd Lts. Not much is expected from new 2nd Lts. So he started working on a field repair.

Date 0530 Monday April 3, 2000
Oklahoma City Air National Guard Base, south ramp
Will Rogers World Airport, OK

Steve met up with Tiffany, Rob and Jacob outside the pax (short for Passenger) area. Jacob was tagging along at the last minute. Lt Jacob McFarland was a graduate in electrical engineering and also was assigned to the C/KC-135 System Program Office. Steve started telling a story about the previous week and Tiffany stopped him and said we have a six hour flight, we can talk on the plane. So everyone agreed to wait for the flight.

Basically flying Space-A is much like a commercial airline except you don't get any snacks unless you brought them, you may have to sit on cargo webbing called "troop seats" instead of a real seat depending on aircraft and the Air Force has zero liability to get you back home in the event the aircraft is re-tasked or broken…so flying Space-A was nothing like flying commercial but it was FREE! Going with a guard unit had a little more stability than an active duty aircraft so they were more likely to get you home and the guard typically performed better maintenance than active duty because they kept the planes and they had pride of ownership.

The TSgt working the desk informed everyone it was time to board the C-130 turbo-prop airplane. Whomever had set this Space-A trip up did a great job of secrecy because there were not a lot of people flying on this chalk to the U.S. Virgin Islands!

The Lts grabbed their bags and walked out the door to the ramp and walked behind the C-130 where the cargo door was lowered and boarded the plane. The troop seats were configured to hold maximum pax so all the troop seats were installed down the outside walls of the fuselage and there was a double row going down the middle of the fuselage so all the seats faced inboard or outboard not forward like standard airline configuration.

Within 30 minutes the C-130 was taxing to the runway. It quickly became apparent there would be no talking. The roar of the four turbo prop engines was deafening. So everyone put on their headphones and listened to music or read books.

Date 1605 Monday April 3, 2000
Henry Rohlsen Airport, St. Croix U.S. Territory

    The C-130 touched down without any issues at the Henry

Rohlsen Airport. All the pax grabbed their bags and headed

out the cargo door. The ramp at the airport was filled up

with all kinds of National Guard aircraft and Steve spied a

C-141 Starlifter from Jackson Ms. He told his buddies he

used to work on that jet.

    One of the flight crew was taking names and phone

numbers so he could get in contact with the group if the

return flight time changed. Then the aircrew announced they

had a bus for anyone going to the group of hotels on the

north side of the island. Steve was like "Hey lets jump on

that bus and get a free ride." Everyone agreed and they

soon were offloaded at the Club Comanche Hotel.

    The group of four Lts didn't bother making

reservations so they went up to the counter to see if they

had any rooms available and they had one with two queen

beds so they took that one. Everything in St. Croix was a

lot cheaper than in the States. They walked to dinner and
decided to get a rental car the next day.

Date 1605 Tuesday April 4, 2000
Strand Street, St. Croix U.S. Territory

They picked up a rental car and Tiffany drove…that was
probably a mistake because they drive on the other side of
the road in St. Croix. Tiffany was an OK driver…on the
right side of the road but not the other side. No kidding
the group was almost killed by the crazy St. Croix drivers
coupled with Tiffany's not great driving on the wrong side
of the road.

Finally, and in one piece they made it to a deserted
beach. Tiffany, Steve and Rob knew each other fairly well
but not so much Jacob. Steve asked Tiffany if she would put
on suntan lotion on his back so he wouldn't burn and
because she was the token girl she did. Steve returned the
favor and then Rob asked her as well.

They had a great time at the beach. No one else was
there. They stayed most of the day and then decided to go
back to the hotel and get ready to eat dinner somewhere
new. The drive back was uneventful.

When they returned to the hotel they tried the key to open the door and it seemed the lock had corroded and was completely broken. They went to the desk and after 30 minutes maintenance final climbed up a ladder to the second floor room window and broke in. Basically they didn't lock the door again that week unless they were inside the room.

After everyone showered they noticed that Jacob's back was lobster red, except for two large pasty handprints on the side of his back where he could reach to put suntan lotion. He failed to get anyone to put sun tan lotion on his back so he earned the nickname "Lobster". Later that night Steve was going to bed early because he had booked a SCUBA dive trip early the next day but his friends were having none of that nonsense. Jacob was throwing towels at him and Rob jumped from his bed over to Steve's bed and started jumping up and down on it until the bed broke. Steve tried to sleep on the sloped bed but it wasn't working so he took out a drawer from the chest of drawers and jammed it under the broken side and that seemed to work well enough.

Date 0805 Wednesday April 5, 2000

Mt. Welcome RD Pier, St. Croix U.S. Territory

Steve had one dive after he went with Mat in the quarry in Alabama. Steve had another trip to Cozumel with some friends in college. Cozumel was awesome with reef diving and night diving.

Steve met the boat captain at slip 12 on the pier and the hot lady dive master came walking up. Steve was the only one going out today so he had a private tour of the "Love Shack" dive area. A quick boat ride to the dive area and the dive master briefed Steve on the dive. He didn't know if he would see any gorgeous reefs today but he did know he would get a pic of the gorgeous dive master.

They back flipped into the water from the boat rail. Then they let the air out of their Bouncy Control Devices (BCD) and descended to 60 feet. At this depth they had roughly 45 minutes under for the first dive and 30 minutes for the second with an hour in between.

There was a lot of sand and some boulders but not much sea life until…they came upon a rock that had a small cave underneath it and Steve saw a huge fin sticking out of the rock. The dive master did not see it until the huge nurse shark swam out, looked at the two divers, turned 180

degrees and swam out of view in the blink of an eye! The
dive master's beautiful eyes were like saucers and so were
Steve's. That was the first shark Steve had seen underwater
and he always thought if a shark came close he would pull
out his dive knife and stab the shark. There is no way to
even get his knife out to try to stab the shark.

The second dive was uneventful and he headed back to
the hotel room around 1400 to discover a very annoyed
maintenance man fixing the broken bed.

Date 1020 Thursday April 6, 2000
Mt. Welcome RD Pier, St. Croix U.S. Territory

All four of the LTs decided to go out on a local
catamaran and snorkeling for the day. This time Jacob asked
for sun tan lotion to be put on his back. There were a
couple of cabin workers on the boat in addition to the
Captain. There was one college girl and then a teen cabin
boy. Everyone teased Tiffany that she liked the young man
but truth be told the men thought the college girl was
fairly attractive too.

They had fun sailing and snorkeling and they even were
able to see a lone barracuda swimming under the boat. Later

when they got back to shore they decided to go shopping

since tonight was their last night on the island.

They went to one tourist trap and they were selling

boatloads of Rum. The stuff was like water down here. What

would cost 12-15 dollars in the states was 2-3 dollars on

the island. The group all knew about customs and that you

would have to pay a lot of money if you went over some

limit but no one knew what the limit was so everyone in the

group bought a couple of bottles of Rum to take back.

Later in the evening they went to a busy

restaurant/bar and ran into some other Americans not from

the island and they started talking. It turns out it was

the crew of the C-141 airplane they saw the first day at

the airport. The crew from Jackson MS. Steve didn't know

any of them personally but they had mutual friends and let

me tell you they were country so much Tiffany started

making fun of them as soon as they left.

Steve worked hard to never take on a Mississippi

accent just because hey it made you sound unintelligent

even if that were not true. One time Steve was interviewing

for a job in Mississippi and the HR person asked where he

was from because he had a generic accent…like a news

caster.

Date 1020 Friday April 7, 2000
Henry Rohlsen Airport, St. Croix U.S. Territory

Time to board the C-130 for the return flight home.
Tiffany went first up the ramp and noticed there were no
troop seats on the centerline of the aircraft. They had
been removed to make way for precious cargo…Cases and cases
of Rum! Suddenly Steve had an epiphany- he did not buy
enough rum.

Date 1945 Wednesday April 13, 2000
OC/ALC Cube Farm

Lt McFarland appeared behind Lt Craken at his desk and
said "What's Up?" Steve turned around to see Jacob still
sporting his lobster tan from the previous week and said,
"You really need some aloe vera gel." "That's not really
funny" said Jacob. "OK", said Steve. Then Jacob said, "I
need a structures engineer to go with me over to the 507
CAMs - the Reserve unit across the ramp. "What for?" asked

Steve. They have a switch in the cockpit they are asking

about being shimmed. "Sure, I can go after lunch." Jacob

said, Meet me at my car at 1300 and we'll drive over." "See

you then."

Date 1315 Wednesday April 13, 2000
507 Air Base Wing, Tinker AFB South Ramp

Sgt McCay met the Lts at the ¾ hangar. The ¾ hangar

was built to house a much small aircraft but retrofitted

with special doors so ¾ of the aircraft is inside the

hangar but only the tail remains outside in the elements.

It is a very strange sight and if the wind is blowing the

aircraft can actually move around inside the hangar so the

nose has to be chained to the ground.

Everyone went up into the flight deck from the ladder.

There was an electrical troop already there to show the

engineers what they had been discussing on email. The

electrician said, "A pilot wrote up a ticket asking about

this switch on the center console. He showed me in the

start-up checklist where it says to flip it to the "on"

position and in the shut-down check list where to flip it

to the "off" position. Lt Craken said, "So what's the

problem?" The electrical troop knew Lt Craken was the

structures engineer because he had worked with Lt McFarland

and knew he was an electrical engineer. He said, "The first Problem is the switch is loose in the panel because this type of switch is designed for a much thicker bulkhead. Our sheet metal shop made a shim that goes around the other switches and raises the switch ¼ inch. We would like permission to use this shim." Lt Craken said, "That is not a problem. I'll sign off on that." Lt McFarland then spoke up, "So why I am I here?" The electrical troop grabbed the switch, unscrewed the lock nut and pulled it out of the console. There were no wire leads attached to the switch making it a dead switch. Lt Craken asked if this was a one off event or if they looked at other aircraft to see if there were wires attached to the switch in other aircraft. "We are three for three – no wires attached or loose wires even close to the switch. When the other deployed aircraft returns we will check them but I suspect this was for a previous avionics suite that has long since been demoded."

Lt Craken spoke up, "So pilots have to turn this switch on and then off per checklist on their check rides and it does absolutely nothing." "That is correct", said Sgt McCay. Lt McFarland said, "I'll see if I can update the Tech order and checklists and at that everyone left the flight deck and returned to their offices.

28.

Wednesday May 10, 2000
Gas station south side Tinker AFB OK

BOOM, "That was a shotgun blast!" exclaimed Major
Baxter while pumping gas. He looked across the street and
saw two EOD HUMVEES in the parking lot of the ABDR Training
Pad. *They must be shooting those old air planes getting
ready to repair them*, he thought. Agnes, the check-out lady
came running out and said to Major Baxter, "Did you hear
that? Did we have a propane tank explode again?" "No, no
the EOD are having fun with their robot again." As he
pointed across the street. Agnes was visibly relieved since

the last time she heard an explosion like that was when

that stupid Airman Basic detonated his home made grill and

sent his little Chihuahua, Fifi soaring into the family

camp pond. It looked like a football going for the extra

point.

SrA Bart Masters had been an Explosives Ordinance

Disposal (EOD) for two years and loved every minute of it.

Currently he still had all of his fingers. Usually Senior

Airman did not get to deploy the $50,300 EOD robot but the

ABDR troops had an exercise coming up and they needed some

damages in the airplanes on the south side of Tinker. All

the senior EOD technicians were deployed or out. The Lt Col

in charge of his shop threatened his life if he damaged the

robot or something else he wasn't supposed to. Retired

airplanes OK, other things/people bad.

The EOD robot was a tracked robot about the size of a

wheel chair. It had four high definition cameras on it and

could see in the visible, infrared and in complete

darkness. It was normally equipped with two articulating

arms with a three finger hand on each but today in addition

to the arms it also had the Remington 1600 automatic

shotgun installed. Typically the shotgun was used to shoot

and trigger Improvised explosive devices down range but not

today. SrA Masters was going to rack up some hits on a B-52

and maybe a C/KC-135.

He met with MSgt Malcolm about twenty minutes ago and

the MSgt showed him where he would like the damages. There

were two per plane. On the C/KC-135 he wanted one on the

wing-to-body fairing on the left wing close to the

fuselage. The second for the 135 would be a shot up through

a balance bay panel and a spoiler. A balance bay panel is a

fairly ingenious way to move flight control surfaces in the

event of hydraulic failure. Basically they use aerodynamic

pressure to move the flight control opposite of the free-

stream air pressure so in the event of hydraulic failure

the pilot could still control the aircraft by means of

heavy forces on the control yoke.

The damages on the B-52 were one to the bottom left

longeron in the forward bomb bay area and then another to

the pressurized area shear panel (body skin) beneath the

cockpit. SrA Master's was stoked. He could make a whole day

out of this outing and not have to do any paperwork or

study boring CDCs for his five level (job proficiency).

He unloaded the robot nicknamed "Major Boom" and

actually maneuvered it to where it fell off the trailer

ramp about a foot or so. MSgt Malcolm watched with wide

eyed skepticism. He wasn't this nervous since he gave a room full of Lts a crash ax. SrA Masters didn't bother giving any kind of explanation and moved the robot in position for the first shot. He took out the checklist and, to his credit, followed it number by number. The last step was to load the shotgun. Before he did so he cleared the entire area of personnel, which was just the MSgt. He was backed up about 20 yards anyway because he didn't trust the SrA after his obvious lacking offloading skills. MSgt Malcolm would have like to use C-4 explosive, as would the EOD guys, but the airplanes were too close to the gas station and there was that whole debacle with the grill and Chihuahua.

The SrA backed away about fifteen yards, yelled "CLEAR". He lined up his shot with the monitor on his controller with the red "X" and hit the fire button. The shotgun did not disappoint. The buck shot tore through the aluminum balance bay and the spoiler. Both men inspected the damage and MSgt Malcolm looked at the SrA and said, "Hit it again". "Yes Sir", said SrA Masters with a huge grin on his face. They had a literal blast. Major Boom did not disappoint. There were some pretty gnarly damages on both aircraft. The B-52 actually had some burned metal from the gun powder and the closeness of Major Boom. Most of the

damages had metal petalling outward from the blasts and

there were some secondary structures damaged as well. Some

aircraft systems had been damaged such as electrical and

hydraulic. "This is some realistic battle damage", MSgt

Malcolm told SrA Masters at the end of the day. "You can

come back next quarter and shoot some more aircraft if you

like." "That would be great. I'll have to find a B-52 and

KC-135 stencil and paint a couple aircraft on Major Boom.

29.

Monday May 15, 2000
ABDR Training pad south side Tinker AFB OK

"Exercise, Exercise, Exercise, RADAR reports three

SCUD missiles inbound with possible chemical, Biological or

Radioactive payloads on board", said the exercise evaluator

Lt Smith. Once a quarter the 654 CLSS would put on an ABDR

exercise at the ABDR training pad. The exercise would be

the culmination of all the individual training classes. The

ABDR technicians and engineers would work together to

install repairs on the pad birds while practicing other

skills such as survive and operate in a chemical

environment, treating wounded with self-aid and buddy care, reporting casualties, force protection, small arms issue and unexploded ordinance (UXO) marking.

Lt Craken was pleased to see everyone acting like this was real world and scrambling for the shelter except the mechanics who were busy covering their tools with tarps so they wouldn't be exposed to chemical weapons and then they made a break for the shelter.

Before Lt Smith called for the mock rocket attack Steve was talking to the two engineers that were in the exercise. Steve also was an Exercise Evaluator. Lt Jones and Lt Roach were both being evaluated. They had both recently been through all the training so they should know how to be efficient ABDR engineers. Lt Craken wanted to make sure they knew what they were going to brief the team when they came in the shelter in a few minutes for the mock attack.

"So what damage are you working on?" Lt Craken asked Carly Roach. "I have the B-52 flap damage and I'm lost on it. The team leader said we may be over run and we may just have to leave it because there is no time for a repair." "What if you gave them permission to fly it out with minimal fuel load, minimum flight crew and no bomb load?

You could also see if the flaps can be pinned so they can't be extended. Then you could instruct them to speed tape over the damage." She was the B-52 engineer that would deploy with that team if there were a contingency. She replied, "That could work." Lt Jones was soaking it all in. She was an engine engineer so wasn't assigned an airframe but she could be put on any team Tinker had. This was her first dealing with airframes.

About that time the team came running in saying missile attack. The team all got their masks on for chem warfare. They already had their suits on because of the threat level. So the team hung out in the shelter for five minutes and Lt Craken gave them the all clear to take off their masks. They still had to wait in the shelter for the other evaluators to plant the inert UXOs outside for the team to find.

TSgt Simmons looked at Lt Roach and asked if she had any ideas on the B-52 damage. She proceeded to tell the team lead all the information Lt Craken and she had discussed earlier. TSgt Simmons was amazed at the solution. Steve was pretending to do work on his laptop but was paying attention to the discussion. Lt Jones was looking at Steve to see if he was going to take any credit for the repair strategy and Steve knew it. He just sat there and

pretended to read email.

Steve had no idea she would deploy with that team

after 9/11. She had built some serious street credit with

that team.

Tuesday May 16, 2000
ABDR Training pad south side Tinker AFB OK

Lt Jones had been working the previous day on the

balance bay and inboard aileron damage. The ailerons on

airplane control the airplane about the "Y" axis or they

control roll. The balance bay was a superficial wound

because it was one of three so if it failed it would have

two other bays to take the load. That became a speed tape

install. The aileron was another matter. Lt Jones,

requested a new aileron from supply once she saw the damage

and received the report. The evaluator said, "There are

zero ailerons in supply." Back to the drawing board. The

assessor said the damage is outside of the Tech order

limits. That means it's up to the engineer. She started her

repair scheme by looking at the Tech order. It was a simple

repair. Flight controls need to be light, stiff and

balanced. The tech order repair had a very thin skin installed over the original structure, which was an aluminum honeycomb with face sheets structurally glued to the honey comb. There was no replacement aluminum honey comb on the trailer so the stiffness would have to come from something fabricated. Lt Jones picked a "T" aluminum extrusion attached to a 0.040 or "forty thousandths" thick sheet of 2024-T3 aluminum and used blind cherry lock rivets to secure the panel on the flight control. She also instructed the ground crew to inspect the panel after every flight to ensure there were no missing or loose fasteners or skin buckling. She figured if the aileron was still attached to the airframe after the first flight that would prove there weren't any flutter issues. See repair write up next page.

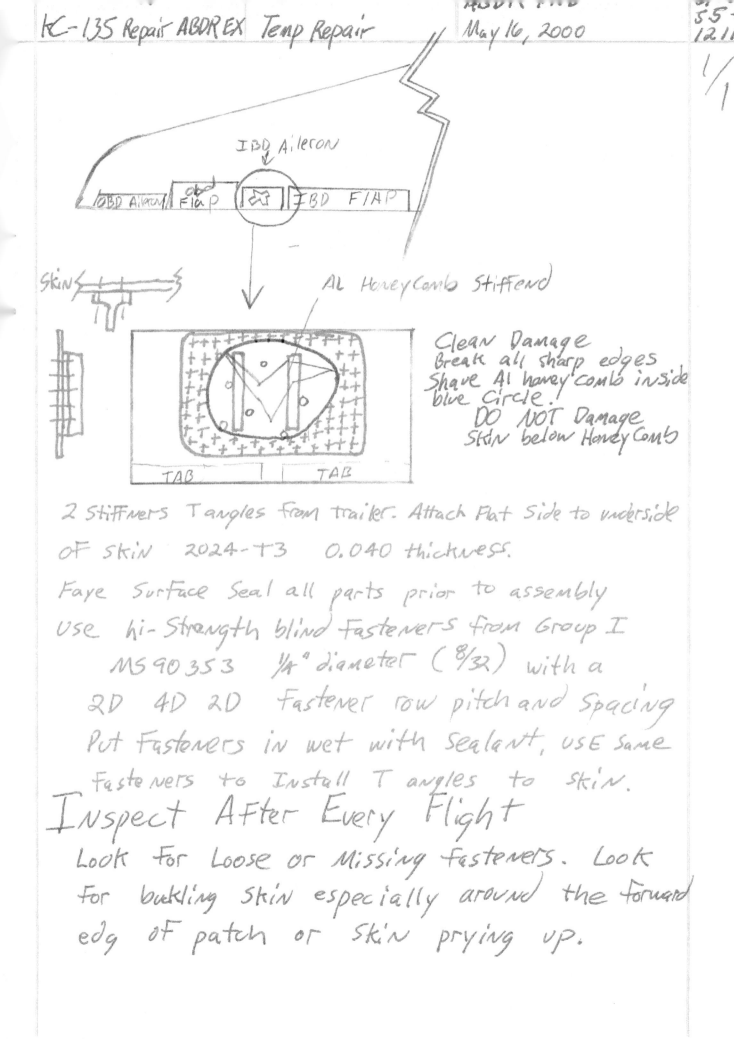

IBD Aileron

OBD Aileroy | OBD Flap | IBD F/AP

Skin

Al HoneyComb Stiffend

Clean Damage
Break all sharp edges
Shave Al honeycomb inside blue circle!
DO NOT Damage Skin below HoneyComb

TAB   TAB

2 Stiffners Tangles from trailer. Attach Flat Side to underside
of skin 2024-T3   0.040 thickness.

Faye Surface Seal all parts prior to assembly
Use hi-Strength blind Fasteners from Group I
    MS 90 353 ¼" diameter (8/32) with a
2D 4D 2D Fastener row pitch and Spacing
Put Fasteners in wet with Sealant, use Same
    Fasteners to Install T angles to skin.

Inspect After Every Flight
    Look For Loose or Missing Fasteners. Look
    For buckling Skin especially around the forward
    edg of patch or Skin prying up.

30.

Friday June 2, 2000
Cube Farm Edgar's Office OC-ALC Tinker AFB

"Lt, there is a jet at Kadena AB Japan with severe corrosion on the right upper vent stringer and wing skin. She is scheduled to come here to depot in eight months. So we can fly her over here to depot and she can sit on our ramp for eight months or sit on their ramp for eight months. Can you go to Kadena and install a temporary repair to keep her flying for eight months?" asked Edgar, Steve's boss. Steve replied, "I definitely can. I will contact the CLSS and see if they can send a team."

Wednesday June 7, 2000
SEATAC International Airport

Steve was on an adventure, like a two or three in a lifetime kind and he realized what it was. He landed on a commercial flight from Will Rogers's international airport to SEATAC. Once he arrived to SEATAC there was a special military terminal for flights to Kadena AB. He found it and checked in. The flight was a commercial flight but contracted to the government so there were only military members or dependents on board. Steve got stuck in a middle row but hey he was going to Japan so he really didn't care.

An hour into the flight he decided he was tired of looking at magazines and decided to try to sleep. He fell asleep and ten minutes later he was awakened by a screaming dependent kid. The kid finally settled down 45 minutes later and Steve went back to sleep. That was a long boring flight.

Lt Craken stepped off the plane in Okinawa and went to clear military security. His contact was waiting for him to pick him up, he saw but the SP asked him, "Do you have anything to declare, Lt?" Steve replied, "No." "Are you sure you don't have anything to declare because we x-rayed

your bag." Steve was tired from his long spring boarded

flight but he finally remembered he packed a diving knife

in his check on luggage. "Are you talking about my diving

knife in my bag?" "Yes do you have your diving license?"

Steve showed the SP his license and dug out his knife. The

SP said, "Really we like to see a knife that has a blunt

tip and not a pointed one but we can let you continue."

"Thanks." Said Steve, gathered his stuff and left. Capt

Sinclair, a friend Steve met at the Aircraft Mishap

Investigation Course, picked him up and took him to the

base car rental lot. In Japan they drive on the opposite

side of the road as they do in America. Steve had to

continuously think about what side of the road he should be

on when driving.

Thursday June 8, 2000
909th Air Maintenance Squadron

    The team was invited to the weekly maintenance meeting

at the 909th. Lt Craken and the team were standing due to

standing room only. The commander looked at Steve and asked

for an update on 62-15632. Steve started giving him his

ABDR spiel that was tailored to providing a "Temporary" repair. The commander asked when he would get his asset back. Steve deferred to the CLSS team lead. They gave a timeline of about two to three weeks because they didn't know the extent of the corrosion. He was less than thrilled but Steve reminded him it was better than the bird sitting on the ramp for eight months.

The team left that exciting waste of time meeting…no donuts and went to the fuels hangar because the 135 had to have the fuel tanks opened because the damage went into the wet bay on the wing.

Lt Craken was shown a desk in the entry way where he could work so he put his pack down and went with the team in the open fuels area to inspect the corrosion. They all climbed up the maintenance stand to get on top of the wing. They went over to the vent stringer at the Wing Station 180 and they saw the white powder on the wing skin and the vent stringer. The white powder indicated aluminum corrosion. Lt Craken instructed the metals tech to remove the corrosion and drew a few circles in yellow china marker on the structure. He also asked him to gauge the depth of the corrosion so he could design the repair according to how much strength was lost.

Friday June 9, 2000
Building 502, Fuel cell hangar

The team met at the North Entry Control Point for the ramp and transferred to the bread truck. The bread truck was a small-blue box truck made for taking tools and parts to the flight line. The 909th let the team borrow it for the duration of their stay. MSgt Bans greeted Lt Craken with a crisp salute at the gate and they both got in the van. In Japan they drive on the other side of the road so at the airport even though the Lt knew this he still tried to get in the driver's side of the cab. The 654th team had a good laugh about it.

Lt Craken knew he was accepted by the group because there were always members that would give him a good ribbing every once in a while. He felt like they thought he was one of their squadron and not an outsider stupid Lt.

The Lt walked into the fuels Cell and greeted the NCOIC of the fuels cell. He overheard them talking about getting some steaks and grilling them the next day. He walked past the offices and met up with the 654th team up on the wing of the bird down for maintenance. A C/KC-135 with

all the fuel tanks open is a spectacle to see.

There are several underwing access ports that are

barely big enough for an undersized human to enter. The

fuel tanks are considered a confined space. If a person

enters the tanks they are required to have undergone

confined space training and have a safety person outside

the tanks to ensure the occupant does not become entrapped

in the tank or lose consciousness. If the person loses

consciousness the safety person can only pull them out

using a rope tied to the occupant, if the structure allows

for it or in the case of -135 wing tanks the person has to

be extricated by the fire department. There is also an

industrial blower blowing fresh air through the tanks 24/7

while the tanks are open.

TSgt Kaufman fist bumped the LT on top of the wing and

said, "Good Morning". The Lt fist bumped him back and

asked, "What do we have?" "The good news is I have cleaned

up all the corrosion by shaving the skin and spar cap.

There was corrosion on the wing skin and on the spar cap

but it did not go down into the webbing of the spar like we

originally thought." Said TSgt Kaufman. "That is good

news." replied the Lt. "What are the depths of the

corrosion?" TSgt Kaufman said, "For the wing skin the corrosion goes all the way through the skin, which is 90 thousandths by about an inch in diameter in three places. The spar cap is only 10 thousands by a half inch in one spot. The Tech order allows up to 20 thousandths to be removed in a two inch area without repair. So I cleaned the corrosion up and there does not need to be a repair on the spar cap.

Aircraft machinists, engineers and structures technicians use thousands of an inch which takes some getting used to. To illustrate 0.5 is a half inch to normal people but to aircraft mechanics and engineers 0.500 is five hundred thousandths of an inch. A quarter of an inch is 0.250 or two hundred fifty thousandths of an inch.

TSgt Kaufman looked worried and made a grunting sound. The Lt looked at him and said, "What's the bad news?" "Well it's these" as he pointed to the row of solid rivets in the area of the corrosion he had removed the day before. The Lt went for the bait, "What about those?" "DEES NUTS!!!" Replied TSgt Kaufman. The Lt just walked over to the maintenance stand and got off the wing shaking his head.

It was time to design the biggest and ugliest scab patch the Lt had ever seen. *How cool is that?* The Lt thought.

Steve walked past the tool room and saw a female SSgt moving a grill around and he was thinking back to the discussion yesterday on the steak lunch. *I need to take the team out for steaks before I leave.* He walked to the entry way desk to start writing up his repair disposition.

The skin in this area is 0.090 of an inch thick or 90 thousandths thick. I'll make the repair 125 thousandths thick in two pieces stacked in a typical wedding cake design, thought Steve. He walked back out to the plane to get some measurements for his design. He walked into the tool room to borrow a tape measure and there was that female NCO grilling steaks inside the tool room next to the fuel cell hangar with an aircraft inside that had open tanks! There were big signs around the hangar that read "Warning open tanks, No fire".

Steve walked out and was thinking, *Should I order her to put the fire out or move the grill outside? Not my monkeys, not my circus, but I need to get away from the tools room!*

Steve completed the repair disposition and gave it to the team. He asked, "Are we working tomorrow since it's Saturday?" The consensus was no because they would not have access to the building. "OK then mandatory formation tonight at 1800 at the Hawaiian Nights Hibachi steak House." Said Steve. "And uniform of the day is Hawaiian shirt". Everyone was making $100 a day for perdiem. Japan was expensive but most of the enlisted guys were eating Vienna Sausages in their rooms so they could bring that money home. Steve was not one of those troops. He was living large and going and seeing what he could on his time off.

Friday 1800 June 9, 2000
Hawaiian Nights Hibachi steak house

Lt Craken waited for everyone outside the restaurant. He hadn't had steak in a while so he was stoked to get some Colby Beef. He heard some of the enlisted members of the team grumbling about having to spend per diem but acted like he didn't hear them. They went inside and sat down. They started looking at the menu and the guys were like I'm going to get some soup or just salad because it was

expensive. The Lt said, "Nope you get a steak. I'm paying."

It was like a traditional Japanese steak house back in the

states but the food was soooo much better. They had a great

time and when they were done Steve said he was going to a

bar across the street to hang out with some members of the

unit they were helping that Steve met in some courses he

had taken. Some of the guys went some went back to their

rooms on base.

Saturday June 10, 2000
Toguchi Beach

Steve had met an enlisted guy in the 909th that was a

certified SCUBA dive master. Steve asked him about diving

and he said he was going on Saturday. "Sign me up", said

Steve. They did a couple of shore dives off Toguchi Beach.

Okinawa has some fabulous dive sights because there is

nothing but corral around the whole island. Getting out

past the corral breakers was difficult. The waves were

strong and it was shallow for fifty yards. They were

snorkeling out but they had to be careful not to get cut on

the corral when they were in the trough of the waves. It

was fairly intense getting out to deeper water but they
finally made it. It was the best dives that Steve had been
on. In fact he was going to give up diving if this dive was
like Florida and nothing to see but sand. There was way
much more. There was a sea snake that was following the
divers but finally left them alone. Then the dive master
pointed out a lion fish in his cave house under a rock. It
was awesome. He was huge with his poisonous tentacles about
the size of a basketball.

They exited after about forty five minutes when Steve
was low on air due to the swim in. They grilled some
burgers for lunch. They didn't go very deep on the dive so
the hour and a half was more than enough time for the OK
from the dive tables. Steve wanted to make sure they were
safe for the second dive. If a diver dives for too long or
takes a second dive too soon they could suffer from
Nitrogen narcosis which is too much nitrogen in the body.
The symptoms are a lethargic drunkenness and can be fatal.
Steve wasn't having any of that and the second dive was
fine. They saw two huge lobsters and the dive master
grabbed one and threw it on the other one but they didn't
fight. One just sat on top of the other one.

Steve was exhausted. He went back to his room and crashed after a shower. He got up around 8 PM got some dinner from Popeye's on base and went back and crashed until Sunday.

He played a round of golf with some of the 654th team…and lost. He wasn't good at golf and really didn't want to be but they invited him so he wanted to go.

Monday June 12, 2000
Building 502, Fuel cell hangar

The team started work at 0900. Lt Craken walked into the Fuel Cell and put his pack down at the desk he was borrowing. He heard the NCOIC of the fuels shop chewing out that female NCO over grilling in the tool room. He had to give her a Letter of Reprimand due to the serious nature of the error. Steve thought, *I should have said something.* He walked into the hangar and got up on the wing. The team had the patches cut to spec, and had drilled out all the rivets securing the skin to the wing tank vent/box beam. Usually the wing stringers, the structure going from the wing root to the tip, had "I" or "T" cross sections (like an "I" beam on a building) but not where the damage was on this bird.

The structure that the wing skin attached to was a box beam, a 4 X 4 inch aluminum beam. It was not solid. It was a square tube. The box beam served two purposes: 1. it was the primary structure the wing skin was riveted to that withstood the wing's bending loads and 2. It served as a fuel vent tube. The fuel vent allowed air to come into the fuel tanks so the fuel could be pumped to the engines. It also allowed for fuel to be dumped overboard in case too much fuel was pumped into the tanks on the ground preventing a rupture of the tank. The tube also allowed the airplane to jettison fuel in flight, in the event of an emergency, to get the aircraft below safe landing weight.

The problem with the box beam is that there was no access to the inside of the beam to buck solid rivets so that meant blind fasteners and in this case, high strength Jo-Bolts along the vent.

Lt Craken worked with TSgt Kaufman bucking solid rivets where they could be installed in the wing skin. He made some smiley faces in some that had to be drilled out and reinstalled. A smiley face in a rivet is where the bucking gun dimpled the rivet. The rivet is no longer structurally sound and must be removed and replaced.

The team worked the rest of the day on installing the repairs.

Friday June 12, 2000
Building 502, Fuel cell hangar

TSgt Kaufman asked the Lt to come look at a problem on the wing repair. The Lt met him on a maintenance stand by the wing tip. TSgt Kaufman pointed to the exposed box beam end and said, "Two of the jo-bolts failed when we installed them and the tails are now loose in the box beam." He knew the Lt would understand the gravity of having steel F.O.D. inside the fuel tank. The pieces could jam up a pump and cause a fire. Lt Craken said, "Oh man that sucks. Maybe we can borrow a borescope and put it down the beam and see if we can grab them or get a big magnet on a rope…", "Whoa whoa, Lt before you go and break your brain I'm just screwing with you!" "Oh thank the good Lord!" Exclaimed Steve. "Actually we are done with the repair and want you to inspect it up on the wing." Steve climbed up on the wing and the repair looked great. It was to spec and…there were no loose jo-bolts. "That looks great", said Steve. TSgt

Kaufman said, Come over here and watch this." They climbed
down off the wing and back to the wing tip maintenance
stand. TSgt Kaufman started pulling a small string that was
going down the box beam. He pulled about eighteen feet out
and tied to the end was a big ball of absorbent material
(like insulation used to soak up jet fuel) and bunch of
metal shavings. The TSgt had pushed the wad down the
"barrel" box beam and ended up pulling out all the F.O.D.
The engineer learned something that day.

Primary Structure Top Wing Skin

Original Material 7075-T6   0.090 thickness

7075-T6 Properties

	Stress (kSI)	
F_{tu}	74	Tensile Ultimate
F_{ty}	66	Tensile Yeild
F_{cy}	65	Compressive Yeild
F_{su}	44	Shear Ultimate
F_{bru}	112	bearing Utimate
E	10,300	Modulos of Elasticity
E_c	10,500	Compression

Repair Material   7075-T6 Clad (Clad is Pure Aluminum on the outside of the Metal used on the Panel to help mitigate Corrosion.)

Two Sheets

0.060

0.120 Total

Original Load Carring Capability

$$\omega_{original} = F_{tu} t \Rightarrow (74000 \tfrac{lb}{in^2})(0.09 in) = 6660 \tfrac{lb}{in}$$

$$\omega_{repair} = F_{tu} t \Rightarrow (74000 \tfrac{lb}{in^2})(0.12 in) = 8880 \tfrac{lb}{in}$$

$$MS = \left(\frac{\omega_{repair}}{\omega_{original}} - 1\right) \times 100 = 33\% \rightarrow Good\ MS$$

$$\frac{L'}{\rho} = \left(\frac{0.75}{\frac{\sqrt{c}}{0.289(0.09)}}\right) \cong 29 \qquad F_c = 55\ kSI$$

Have to check inter fastener buckling

Use slenderness Ratio $\dfrac{L'}{\rho} = \left(\dfrac{\frac{S}{\sqrt{c}}}{0.289\ t} \right)$ S = fastener pitch
fastener pitch = 4D

C = boundary condition

$\dfrac{L'}{\rho} = \left(\dfrac{\frac{0.75}{\sqrt{1}}}{0.289(0.12)} \right) = 21\ (\text{for } t = 0.12)$

$= 43\ (\text{for } t = 0.06)$ t = skin thickness

use ¼A solid in skin

and ¼A Jobolts in spar cap

$2(.09) = ?$

$A_{original} = 0.18\ in^2$

$F_c = 46\ KSI$ or $61\ KSI$
from $\dfrac{L'}{\rho}$ graph $t = 0.06$ $t = 0.12$

$Area_{original} = 0.09(0.75) = .0$

Spar Cap width = 2 in $Area = 0.12(0.75) = 0.09$

$0.06(0.75) = 0.045\ in$

$\dfrac{P_c}{0.06} = (46\ KSI)(0.045\ in^2) = \boxed{2070\ lbs}$ $\to$ Inter Fastener buckling

$\dfrac{P_c}{0.12} = (61\ KSI)(0.09\ in^2) = 4950\ lbs$

$\dfrac{P_c}{original} = (55\ KSI)(0.0675\ in^2) = \boxed{3712\ lbs}$

Make the bottom 0.120 thick piece Cover all
Damage, to avoid inter fastener buckling.

Fasteners   NAS 1669 (Jo-bolt)

 Diameter 3/, in   Sheet $t = 0.120$ 2620 lb

$P_{su} = 4500$ $F_{bru} = 148 KSI$

$P_{bru} =$ (3820)   Lower of the two

 Number of rows $= \dfrac{\omega_{original}}{P_{bru}} = \dfrac{3712 \text{ lbs}}{3820 \text{ lbs}} = 2$

$$0.97 \sim 1 \text{ Row}$$

Need one Row Jo bolts in Spar Cap + Skin
then   Solid button head rivets every whare
els.

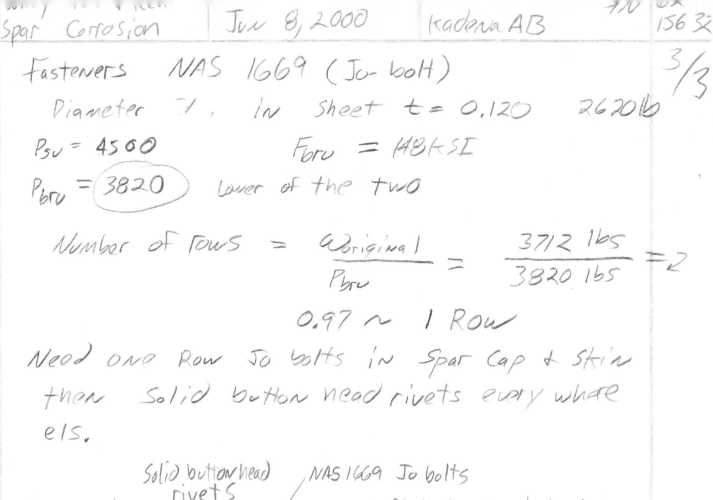

Solid button head rivets   NAS 1669 Jo bolts
Solid button head rivets
Two repair patches
original Skin   0.090
Vent Stringer box beam

Put in fasteners with faye Surface Sealant and between
repair and original Skin. Break all Sharp edges
and chamfir Leading and trailing edges.

31.

Monday October 18, 2000
MacDill AFB Tampa Bay Fl.

   C/KC-135 Tail number 59-1873 was sitting on the ramp after a short deployment to the sand box to support some Air National Guard A-10 units doing training. TSgt Hurls was working on the post deployment short inspection. He had crawled up into the right main landing gear wheel well to inspect all the primary structure there.

   In the wheel well of a 135 there are several things going on; One, that is where the main landing gear truck is stowed when the gear are retracted in flight, two all the

aircraft structure loads and payloads are transferred to
the ground while there is weight on wheels and finally that
is where the 135 is refueled on the ground with a single
point refueling valve. Before this single point valve was
installed a fuels technician had to go up on the left and
right wings and put fuel in each of the wing tanks and then
transfer fuel to the body tanks. There are also various
hydraulic lines and pumps going through that area. An
interesting note is that the Air Force always paints
internal bays of aircraft gloss white so the pilots and
ground crew can see a hydraulic leak easily. Hydraulic
fluid is red so it shows up as a high contrast to the
white.

TSgt Hurls had his flash lite and mirror and started
the inspection of the main landing gear trunnion rib. The
MLG rib channeled all the loads from the wing and
transferred them to the MLG truck. It was a huge steel "I"
beam. He noticed some red flakes on the rib and cleaned the
area with a rag and some came off on the rag…"Rust", Said
the TSgt.

He returned to the shop and started diving into the
Tech Order. His boss SMSgt Ford came up, "Hey Sam how does
1873 look?" "Overall not bad but the MLG Trunnion has some

rust on it." "Oh that's not good. I don't think there are

any allowables on that part for repair." TSGT Hurls said,

"Looks like you are right." I'll have to call Tinker after

lunch." SMSgt Ford replied, "I hear they have a sharp Lt."

"Are you talking about Lt Craken?" "Yes, I think that is

his name." "Right I have worked with him before. Apparently

there is one."

Monday 1300 October 18, 1999
Tinker AFB Cube Farm C/KC-135 land

(In official voice) "KC-135 this is Lt Craken". "Sir

this is TSgt Hurls at the 6 ARW (Ariel Refueling Wing)".

The Lt asked, "How are you today TSgt?" "Good", came the

reply. I have 59-1873 back from a deployment with a small

rust spot on the right MLG Trunnion Rib." "You are going to

have to remove and replace it. There is no allowable on

that part. It is critical." "Right, but could you come and

look at it to make sure we have to R square it?" Steve

asked, "Where are you located?" there were several bases

with C/KC-135s and Steve hadn't bothered memorizing all the

unit numbers. "We are stationed at Tampa Bay Fl." "I will

definitely check with my boss". "Thank you" "I will let you

know", said Steve.

Lt Craken went to his boss and explained about the corrosion on the MLG rib at MacDill AFB. Steve was still relatively new so his boss wanted him to get experience and the program office had problems in the past with engineers ignoring their problems or taking a really long time to answer a field call. So Edgar, his boss agreed for him to go Temporary Duty Yonder or TDY to Florida.

Steve was stoked he went back and started the paperwork for the trip and sent TSgt Hurls an email telling him he would be there Thursday.

He made plans to fly out Wednesday, work Thursday at MacDill and fly home Friday.

Steve had been to Florida a few times with his family and once when he was in high school on a senior trip. He had never been to Tampa so he was fairly stoked to get a work trip there.

Wednesday 1530 October 20, 1999
Embassy Suites Downtown Tampa Bay FL.

Steve arrived at his hotel downtown and checked in. He could have stayed closer to MacDill AFB but he wanted the whole Tampa experience. He went to a nice restaurant and went driving around the city with the culmination of the

sunshine bay bridge.

Thursday 0830 October 21, 1999
Embassy Suites Downtown Tampa Bay FL.

    Lt Craken drove his rental car, an upgrade from what

the clerk told him was a Chrysler Town and Country

minivan…*How is this an upgrade?*, thought Steve to the front

gate at MacDill AFB. He drove around to building 860, the

Structures generation flight and walked in looking for the

structures supervisor. He found SMSgt Ford and introduced

him. SMSgt Ford thanked him for coming and set up a trip to

the flight line. The Mechanic assigned to this bird was

TSgt Hurls and he took the good Lt out to the plane and

handed him a flashlight, mirror and a scribe. The Lt

crawled up into the wheel well and scratched around on the

MLG Trunnion rib. "Sorry you are going to have to remove

and replace it. It has scaly rust about an inch wide.

    The TSgt and crew were grateful Steve had come out and

invited him out to Cowboys grill for lunch. Steve accepted

and they were all talking about a waitress named Darlene

and how they hopped she was working today. Steve picked up

on the joke but played dumb and went along.

Cowboys was a local tradition along the lines of Hooters. TSgt Hurls came in first and asked for Darlene. She was working today. Darlene enjoyed giving the Airmen special attention and in return got high tips. She was enough to make Steve blush along with her antics.

After lunch Steve went his separate way. He calculated that he had worked a grand total of 45 minutes. He headed back downtown to the hotel to change and then went to Busch Gardens. It happened to be military appreciation week so Steve got in free and received coupons for a meal in the park and there was hardly anyone there. Steve rode the rides until he was almost ready to puke. What a great trip. He also had plans for a dinner with a friend he met at Church but moved to Florida to manage a hotel. What a great trip.

32.

Date 0330 Sunday April 1, 2001
Approximately 300 feet deep in the Pacific Ocean 450
Nautical Miles SW of Taiwan

The USS Memphis SSN-691 was hovering at 2/3 standards

speed right above the thermal cline at 300 feet deep. The

thermal cline is where the temperature rapidly decreases in

an ocean. This phenomena lends itself to be able to "play"

with an enemy's SONAR. SONAR (Sounds navigation and

Ranging) is similar to RADAR (Radio Detection and Ranging)

in the air. If a sub is threatened it can descend through

the thermal cline and decrease speed which decreases the

subs sound and possibility of detection or tracking. Some

SONAR from surface vessels can bounce up off the thermal cline and the sub underneath is virtually invisible.

Captain Jeffery and crew had finished up war gaming with Japan and were near the end of their 6 month tour. Their home base was Naval Base Point Loma in Southern California. If the boat headed for home now it would be a little soon for the new sailors to get their long deployment medals so the subs XO (executive officer-second in command) suggested a cruise over toward Vietnam to see if they could get some SONAR tracks of the Vietnamese Kilo class submarines that typically patrolled close to their territorial waters.

The Vietnamese Navy called Binh chung Tau ngam had up to six Russian designed and built Kilo class submarines that they operated. The Kilo class submarine is one of the quietest class submarines produced because it has battery propulsion but the downside is it has to recharge the batteries at certain intervals by snorkeling and running its diesel engines.

The Captain agreed and addressed the Helmsman to come to a heading that would lead to Vietnam.

Date 0500 Sunday April 1, 2001
Sub Pins 6 and 7 Hainan Island, China

Major Wong was in his office early this Sunday Morning

due to the maiden launch of code name Kuái Shâ yĩ háo and

Kuái Shâ yĩ ěr or "Fast Shark One" and "Fast Shark Two".

The two super subs were finally ready to launch a little

behind schedule due to Mother Russia dragging their feet on

delivery of the second sub and the accident in the old sub

pin area involving the "Sampson" sea launched cruise

missile. It had taken nearly two years to repair the damage

the dropped munition had done to the subs outer hull.

The accident was ruled by the Chinese investigators as

sabotage by the Russian "Advisors" instead of a black eye

to the Chinese Navy for not properly securing the missile

to the crane rigging. One detail missing from the report

was there were no Russian Technicians allowed in the sub

pin.

The two nuclear subs were nicknamed Kuái 1 and Kuái 2.

The subs were almost identical except for nuances only

known by the Russian designers. Both subs were fully

stocked with food and munitions and were ready for their

deployments to the Pacific.

Kuái 1 was ordered to penetrate Japan's coastal

waters and perform Intelligence, Surveillance and

Reconnaissance or ISR of the joint bases with the

Americans. Kuái 1 was second out of the gate around 0900.

Kuái 2 was ordered to perform the same ISR mission to

the Indian Ocean to watch the nations of Pakistan and

India. Kuái 2 was first out of the gate around 0830. This

sub also had a small mini sub attached to the egress trunk

on top and a compliment of four Chinese commandos to work

on insertion operations.

When World War II ended the victors drew up Geo-

political lines and really messed up the political climate

of the globe for years to come. You can see evidence of

this especially in Africa and the tiny mountainous area of

Kashmir. Kashmir is a beautiful mountainous area that is

cool in the summer and fairly mild in the winter. So

basically that region is a gorgeous oasis amongst the

armpit of the hot steamy world. Three nations want that

area and thanks to the British that drew up the map, they

didn't bother to put lines in that mountainous area because

hey, who would fight over very high scraggy mountains, well

China, India and Pakistan, that's who.

China built an all-weather road into the mountains for

the sole purpose of attacking India and Pakistan and they

did it secretly while India was fighting with Pakistan and

then one day China fired on India and Pakistan from a

higher overlooking position. "How did they get artillery up

on the mountain?" the Indians asked. Well they built a road

up the backside while you were fighting Pakistan.

Kuái 2 headed south to perform ISR enroute on the

Malaysian navy and do a little war gaming on them without

them knowing. Their main mission was ISR on India and

Pakistan while orbiting in the Indian Ocean.

Date 1200 Sunday March 25, 2001
CIA Headquarters Langley, Virginia
A week previous to the two Chinese subs launching.

The CIA had noticed some interesting HUMINT coming out

of China. HUMINT stands for Human Intelligence or in

laymen's terms spies. The CIA had been recruiting Chinese

assets from new immigrants to the U.S. and the operation

CATFISH started paying dividends. One such asset fled

communist China because two of her family members were
killed protesting during Tiananmen Square. Her boyfriend
was also missing. Lao's boyfriend name was Wong and he was
a dock worker at a top secret sub base possibly on Hainan
Island. Lao said, "Wong was worried about an accident that
killed several other workers loading something on a weird
looking submarine. When I did not hear from him in a week I
escaped to Taiwan and boarded a ship to the U.S."

The CIA handler contacted his Navy liaison at the
Pentagon and told him to check his STU III (a secure phone)
fax machine for intel on China.

Date 0700 Monday March 26, 2001
Pentagon, Arlington, Virginia

Navy Captain Pierce checked his voice mail and then
got the keys for the STU III. He really hated that CIA dirt
bag but dealing with the CIA was better than getting some
general officer or admiral his morning cup of coffee, which
reminded him to start some brew going ASAP. Then he went to
the vault where the STU III was kept. He swiped his badge
and entered his pin and the door unlocked. He walked
through and closed and it automatically locked back. He

entered the key into the console and hit the receive

button. The first fax was some spam on a time share in

Fiji. He threw that away and wondered how in the hell did a

time share from Fiji get on the classified network? That is

a job for the IT guys. Then his CIA fax came through.

Captain Pierce read through all the HUMINT and was vaguely

interested. Some hinting that the Chinese might have

themselves a big boy nuclear sub. That could be a game

changer.

Captain Pierce would meet with PACOM (Pacific Command)

later today to see if they could get some ISR around the

island that the HUMINT suggested contained a secret sub

base.

Date 0930 Sunday April 1, 2001
41,000 feet over the Pacific Ocean 150 miles east from
Hainan Island, China

Aircraft Commander Kyle (Smitty) Smithey asked the co-

pilot to open the orders for the day. LT Gray went aft to

the aircraft safe and entered his key and waved to the

senior enlisted SIGINT (Signals Intelligence) operator

Chief Petty Officer Landers to come to the safe and insert

his key to retrieve the orders. Landers came over taking

his sweet time because Lt Gray had won $259 dollars and 52

cents from him last weekend at the joint Enlisted/Officer's

club. Enlisted and officers aren't supposed to gamble but

it happens. When the safe was opened the first set of

orders were a red and white Emergency Action Message (EAM)

encased in plastic. Landers looked at Lt Gray and said,

"It's going to be one of those days". The Lt grabbed it,

broke it opened and ran to the flight deck. He got back in

his seat, strapped in and put his headset on and told the

commander he had a properly formatted EAM. "What the hell

is that?" The commander had never seen one before. Usually

only submarines got EAMs. "What does it say?" He asked the

Lt. "We are to proceed to 40 miles within Chinese

territorial waters close to Hainan Island and listen for

any evidence of a new super submarine." "Come to new

heading 270 and Angels 32 for mission profile", said the

commander to the pilot. The back-seaters got word to format

their sensitive equipment for submarine detection and

classification for a new enemy sub.

Date 0945 Sunday April 1, 2001
Long Range Defense RADAR site Chi, Hainan Island, China

Petty Officer First class Chin was manning his RADAR console and noticed a possible contact on the fringe of his scope to the east. He alerted his supervisor and his supervisor told him it was probably noise or a solar flare. "Check the space weather report", he ordered Chin. Chin got up from his console and checked the binders for space weather on the shelf. The binders had not been updated in two days. Chin's best friend was responsible for updating the binders but it was one of 50 additional duties. The supervisor asked if there were any solar flares released in the last day. Chin covered for his friend and said, "Yes two solar flares erupted in the past 24 hours." "That settles it then, no contact". Unfortunately for Chin when he got back to his console there was definitely a new contact from the east. "SUPERVISOR, CONTACT BEARING 090, ALTITUDE 9,750 METERS, VELOCITY 350 KNOTS!" "Designate contact alpha and run the contact checklist", instructed the supervisor. Chin ran the checklist and the first step was to contact the owner of the Military Operating Area (MOA) to see if this was a friendly aircraft even though there were no notices of friendly forces training today. With that completed and no answer on the phone from the MOA owner, which was a poorly led Chinese reserve unit, the next step was to hit the fighter scramble button. Chin

looked at the supervisor and he nodded to hit it. It took

30 minutes from the first contact to hit the Claxton

button.

U.S. Aircraft P3 Sub Hunter ISR gathering aircraft Call
Sign Black Sheep 3
Date 1000 Sunday April 1, 2001
32,000 feet over the Pacific Ocean 60 miles east from
Hainan Island, China

Ensign Rands, the P3s Patrol Plane Tactical

Coordinator ordered the offensive enlisted member, Petty

Officer Blake, to drop a SONAR buoy. They were still a ways

out but wanted to get one in the water to start a baseline

for future buoys they would drop to detect any new Chinese

subs. The baseline for the buoy consisted of salinity

content, water temperature, atmospheric temperature and

pressure.

The buoy launched from the rear of the plane and hit

the water a few minutes later. The buoy detected the

landing in the ocean and started broadcasting soon after

that to the P3. It went through its initialization and

within one minute was up and registering ocean sounds.

Interesting enough the buoy detected a contact. "Contact, designate ECHO bearing from 045 - 090 degrees, range approximately 55 nautical miles from buoy. Signature software running…Contact Echo is submerged and very faint. Echo is probably a submarine. Echo is confirmed a Los Angeles class U.S. submarine." Petty Officer Blake asked the Ensign to notify the aircraft commander there was a friendly in the area. The Ensign said, "Roger that", and walked up to the flight deck and alerted the commander. The commander said, "Drop another buoy and see if we can triangulate." "Yes Sir", the Ensign said as he went back to the crew compartment. The Ensign gave the order to drop buoy number 2 and they did. It started broadcasting soon after hitting the water.

U.S.S. Memphis Submarine
Date 1015 Sunday April 1, 2001
Approximately 300 feet deep in the Pacific Ocean 500 Nautical Miles SW of Taiwan

"CONN SONAR, SPLASHES IN THE WATER"! Captain Jeffery was in his quarters after the late wrap up of the exercise. He answered the SONAR station call and headed to the bridge. "What do we have SONAR?" "We had one splash 3 minutes ago and now just had another one approximately 40

miles west of the original splash position making the

platform an aircraft probably a turbojet airplane dropping

SONAR buoys." The Captain ordered, "Helmsman make your

depth 450 feet 10 degree down bubble, come left to new

heading 200 degrees. Chief of the boat sound battle

stations!" "Aye aye Captain, sounding Battle Stations",

exclaimed the Chief of the boat. The Captain said, "The

plane is probably ours but Russia has been known to fly

this far south."

U.S. Aircraft P3 Sub Hunter ISR gathering aircraft Call
Sign Black Sheep 3
Date 1015 Sunday April 1, 2001
32,000 feet over the Pacific Ocean 60 miles east from
Hainan Island, China

"New Contact, designate Foxtrot bearing 270 degrees,

range approximately 100 nautical miles from buoy. Signature

software running…Contact foxtrot is submerged and very very

faint. Foxtrot is probably another submarine or a whale

eating a shark. The software does not have this contact

previously cataloged." Petty Officer Blake said, "This is

a new class of submarine." Ensign Rands started recording

the track of the possible new sub. "Commander, we have a

possible new submarine that is not friendly. Do you want me
to send a satellite flash message to Pacific Fleet
Command?" asked the Ensign. "Yes and see if there are any
TACAMO aircraft up that can talk to our sub. If there are,
have them send an alert to our sub about the new Foe in our
vicinity." said the commander. Petty Officer Blake walked
over to the Comms Specialist and relayed the info about the
TACAMO request.

Take Charge and Move Out (TACAMO) is the secure
communication system that the U.S. uses to talk to the
nuclear Triad (land based intercontinental ballistic
nuclear missiles, air launched nuclear missiles and
submarine launched ballistic missiles). The Navy shared
part of the Air Force Base at Tinker on the south ramp. The
unit there has eight Boeing 707s that are very closely
related to the C/KC-135. They have been highly modified
over the years in service. Their designation is E6-B
Mercury and they are painted anti-flash white in contrast
to the all grey C/KC-135s. The TACAMO mission for the E6-B
is to talk to the U.S. Subs using Extremely Low Frequency
(ELF). ELF frequencies are on the order of 3-30 Hertz or a
length up to 10,000 kilometers! So basically the wave

lengths are ginormous which allows them to penetrate the
ocean.

By a total act of random schedules there was one
TACAMO E6-B flying training close to the Japanese coast at
Okinawa. White Sheep 4 was flying out to a predetermined
location to perform training with a fairly new crew. The
Black Sheep 3 crew made contact and relayed the message
they wanted the U.S. Sub to know there was a new enemy sub
in the area of operation. The aircraft commander, Commander
Rodriguez ordered the crew to start paying out the five
mile long antennae out the back of the jet. The plane flies
in a constant level right bank paying out the five mile
wire so that it becomes a helical antenna that is 3000 feet
long!

There was an incident with one E-6B that the antenna
became fouled and they could not reel it back up so the
aircraft commander gave the order to cut it and it ended up
in some farmer's field. The Navy had to pay a lot of money
for the "premium crop field".

The Communications officer in the back of the jet let
the commander know the antenna was fully deployed thirty
minutes later. Then he started working on the Emergency
Action Message (EAM) they were going to send to the U.S.

Sub which they had a good idea it was the U.S.S. Memphis

since they just finished up war gaming with her.

Kuái 1 People's Republic of China Submarine
Date 1015 Sunday April 1, 2001
Approximately 30 meters in the Pacific Ocean 20 Nautical
Miles east of Hainan Island

"CONN SONAR, SPLASHES IN THE WATER"! Captain Wing Ping

of the Kuái 1 was on the bridge. He walked over to the

SONAR station and looked at the SONAR display. He asked,

"Sonar Buoys?" The operator confirmed that he thought they

were. "Are they ours? There is no way the Americans know we

exist." "Unknown, sir", Replied the SONAR operator. "What

is the bearing?" the commander asked. The operator looked

at the tracks and estimated the bearing to be 30 to 90

degrees and 100 to 130 kilometers range. Captain Wing Ping

grabbed the intercom and announced, "Battle Stations,

Battle Stations" and ran back to the bridge. "Helmsman dive

to 130 meters 20 degree down bubble turn left come to new

bearing 045 degrees.

Kuái 2 People's Republic of China Submarine
Date 1015 Sunday April 1, 2001
Approximately 45 meters in the Pacific Ocean 40 Nautical
Miles south of Hainan Island

"CONN SONAR, SPLASHES IN THE WATER"! Captain Zhang of

the Kuái 2 was on the bridge like his peer on the Kuái 1.

"This close to home?" Captain Zhang said. He was thinking

this must be a test from high command. He stated, "This

must be part of the sea trials. Make your depth 200 meters

20 degree down bubble, come right heading 200 degrees make

revolutions for flank speed." Captain Zhang grabbed the

intercom and announced, "Battle Stations."

U.S.S Memphis Submarine
Date 1020 Sunday April 1, 2001
Approximately 450 feet deep in the Pacific Ocean 520
Nautical Miles SW of Taiwan

The subs' Executive Officer, XO for short was checking

all compartments ensuring they were ready for battle and

his final stop was the radio console. Just as the XO

arrived the EAM light and buzzer sounded. "Well this is

getting interesting!" the Chief of the boat met the XO at

the radio console and read the EAM. "XO, we have a properly

formatted EAM." The XO replied, "Agreed we have a properly

formatted EAM. Open the safe and grab cookie number 12."

The EAM system was an Extremely Low Frequency (ELF) radio

that could be broadcast by the TACAMO in the open ocean for

hundreds of miles and the sub could pick it up while

submerged. Then the crew would authenticate it because

anyone could have broadcast that message and try to spoof

Pacific Fleet Command. Cookie number 12 was removed from

the safe and broken open. The cookie was a hard plastic

case that held the message. The message stated,

------------------Contact requested by US aircraft in your

vicinity. Surface or float your radio buoy to communicate.

END OF MESSAGE---

The XO said, "Captain we have a request to contact

U.S. aircraft. That is probably who dropped the buoys." The

captain replied, "Make it happen, come up to a depth of 200

feet and float the radio buoy. The XO said, aye aye

Captain, that means the aircraft probably detected us."

U.S. Aircraft P3 Sub Hunter ISR gathering aircraft Call
Sign Black Sheep 3
Date 1028 Sunday April 1, 2001
25,000 feet over the Pacific Ocean 60 miles east from
Hainan Island, China

"Sir, we have the U.S.S. Memphis sub on the radio."
said the P3's radio Operator. Captain Smitty said, "Tell
them of possible new sub contact we picked up and give them
a bearing to it. We will remain on station another two
hours performing ISR." "Yes sir". The radio operator
broadcasted this message, "U.S.S. Memphis, Black Sheep 3,
there is a new sub contact to your West that has never been
catalogued, I will send the sound bite following this
broadcast and we are on station another two hours. How
copy?" "Black Sheep 3 U.S.S Memphis copies new enemy sub
west of our position and black Sheep 3 on station for two
hours. U.S.S. Memphis requests Black Sheep 3 run
interference for us to locate and track bogey without
detection." The radio operator asked the aircraft commander
the Memphis request and the got the affirmative. "U.S.S.
Memphis, Black Sheep 3 will run interdiction with the
bogey." "Copy all Black Sheep 3, U.S.S. Memphis out."

U.S.S. Memphis Submarine
Date 1030 Sunday April 1, 2001
Approximately 200 feet deep in the Pacific Ocean 377
Nautical Miles SW of Taiwan

"Chief of the boat, stow the radio Buoy. Helmsman make
your depth 450 feet 10 degree down bubble, all ahead flank
and come to new heading 270 degrees." The Helmsmen
correctly repeated the Commander's order. "We have new
orders, said the captain. "Apparently we have a new Chinese
foe to address. SONAR start looking for this sneaky
Chicom." "Aye aye Captain, said Seaman Halls.

Date 1030 Sunday April 1, 2001
Base operations, Hainan Island, China

The Claxton sounded and Mahjong tiles fell to the
ground. Flight crews of the Peoples Liberation Army Navy
(PLAN) run to their J-8II fighters, ground crews scurry
around removing the "remove before flight" ribbons and
getting the fighters ready to launch. Major Li was jogging
to his alert bird thinking this is another exercise until
he saw the ground crew loading the new air to air missiles
onto his J8. Ok this is real. Lt Yang did not look like he
was taking the alert Claxton serious either until the Major

yelled, "This is real. Get your crap up tight". *Oh, I'm on
it the Lt thought*. The pair jumped into their fighters,
saluted smartly and taxied to the runway and hit after
burners while they were still lining up with the
centerline. The rest of their flight was right behind them.
In all, a four ship of China's elite fighters took off from
the strip at Hainan Island.

Major Li took the lead due to his experience and Lt
Yang was his wingman. The controller gave them vectors to
the bogey. The controller also relayed information on speed
of the bogey. Major Li deduced it was an American Sub
hunter/ISR bird they were to intercept that day. He loved
thrashing the P3 crews. He could really do whatever he
wanted to the bird except shoot it down. In months previous
he had harassed three P3 crews being very aggressive toward
them and flying way too close. He was invincible…but he had
never met an F22 Raptor.

Chinese High command broke the normal radio chatter
requesting to talk to the intercept pilot directly. This
was very strange to Major LI. Major Li said, "Go ahead for
intercept pilot". High command instructed, "Intercept the
bogey, come along side and make visual contact and then
break contact and force it down to land at Hainan Island

using any and all measures except shooting at the bogey.

Make it look like an accident. DO YOU UNDERSTAND YOU CANNOT

SHOOT ANY MISSILES OR GUNS AT THE BOGEY?" Maj Li confirmed

the new orders.

U.S. Aircraft P3 Sub Hunter ISR gathering aircraft Call
Sign Black Sheep 3
Date 1045 Sunday April 1, 2001
20,000 feet over the Pacific Ocean 52 miles south east from
Hainan Island, China

    The P3 copilot saw the flight of four J8s headed

toward them off his right wing. He was used to the Chinese

harassing them in international waters. China declared

their territorial waters extended out to 60 nautical miles

off their shore but international treaties declares only 20

nautical miles. The copilot alerted the airplane commander

about the bogies.

J8II People's Republic of China Fighter Aircraft
Date 1105 Sunday April 1, 2001
20,000 feet over the Pacific Ocean 62 miles south east from
Hainan Island, China

    Maj Li instructed the flight to slow down and cover

him. From what no-one knew because the P3 had no offensive

weapons. He then pressed the throttles up and made a pass

at the P3 at over 500 knots. He came back around and made

another pass. The P3 crew did not alter their flight plan.

Maj Li did hear the P3 commander trying to hail his flight.

"This is U.S. Navy flight 410 in international waters.

Cease and desists all actions. We are legally flying in

international waters. You are being aggressive and

dangerous. We will request fighter escort if you persist."

He did not respond. On the third pass he knew what he

needed to do.

Maj li slowed to match the P3 speed and brought his

fighter alongside. He used hand jesters to tell the P3 to

depart the area. The P3 crew ignored him. He then

accelerated off and turned back to make a high speed pass.

Coming close to the P3 he edged his fighter too close and

clipped off the left wing tip of the P3. When he clipped

the P3 wing it also tore off a large portion of his right

wing and the majority of his fuel. His flight computer went

blank and then 1.5 seconds later the master alarm sounded

and he quickly decided to bail out. He pulled the ejection

handle and departed violently from the condemned J8. He

cleared the jet and separated from his seat but his chute

did not open and he plummeted 15 thousand feet into the

pacific.

U.S. Aircraft P3 Sub Hunter ISR gathering aircraft Call
Sign Black Sheep 3
Date 1045 Sunday April 1, 2001
20,000 feet over the Pacific Ocean 52 miles south east from
Hainan Island, China

"THAT CRAZY SOB JUST COLLIDED WITH US!" Exclaimed the

aircraft commander as multiple alarms started going off in

the cockpit. "Go through the systems and start silencing

the alarms one by one so we know what we are dealing with"

stated the commander.

The Copilot said, "We are dealing with a hydraulic

problem for sure. Flight controls are sluggish and I'm

putting in max roll input to the right to stay level. I'm

slowing down and decreasing altitude." "Make it happen and

turn on emergency hydraulic pumps" the commander calmly

said.

The commander yelled to the back seaters, "Hey someone

look for a chute on the J8. I think it broke up." "Aye sir

it did break up with the pilot ejecting but no chute."

replied Ensign Rands. The commander looked at the copilot

and asked, "Can we make it back to Okinawa?"

"No sir. Looks like we are leaking fuel and we have to fly slow." "Copy" replied the commander. "Pacific Command, Black sheep 3, May Day, May Day, May Day, a Chicom (Chinese Communist) J8 sheared off our left wingtip and we are leaking fuel and hydraulic oil. Cannot make it back to base. Our options are to ditch in the ocean or land Hainan Island PRC." Pacific Command replied, "Black sheep 3, Pacific Command "Aircraft Commander's discretion on course of action. This is duty officer of the day authorizing *Broken Trident protocols,* how copy?" The aircraft commander said, "Acknowledged Broken Trident protocols."

Lt Commander Smithey looked at the copilot and asked him, "Do you feel like swimming today?" An Emphatic "NO" was the answer. "Very well set a course for Hainan Island."

Lt Commander Smithey asked Ensign Rands to come up to the cockpit on the intercom but no response. The aircraft commander took his headset off and yelled, "Ensign Rands get up here." Ensign Rands ran to the cockpit. The aircraft commander looked him in the eye and gave him the Broken Trident order. "Do you understand? The commander asked. "Yes sir," was the reply.

Ensign Rands ran back to the mission area of the plane and gave the Broken Trident order. All the mission specialist opened their emergency binders and started running the broken Trident checklist for their individual stations.

The ensign ran over to the safe and entered the combination and opened it. He got back on the intercom and requested to open the crew door to start jettisoning classified equipment into the ocean. He got the go ahead and warned the rest of the crew he was going to open the emergency over wing hatch to jettison equipment.

The Ensign ordered the Petty Officer to man the crash ax to destroy electronics and then to throw it overboard. Petty Officer Karls grabbed the ax and started wielding it like Jack Torrance in the movie "The Shining". Other crew members stopped what they were doing and just watched the carnage on the electronic equipment they were just using five minutes ago which constituted millions of dollars of highly classified equipment.

Ensign Rands went to each station collecting classified paper and decided to burn it in the safe. He started a fire with is Zippo lighter before thinking that was a bad idea. Just after that the aircraft commander yelled back, "I smell smoke are we on fire?" The Ensign yelled "NO" without explanation and grabbed the crash ax from Karls and pounded the aircraft structure the safe used

to be attached to, closed the door to the safe and locked

it. He grabbed Petty Officer Karls and yelled, grab the

safe. We are going to throw it overboard!"

Kuái 1 People's Republic of China Submarine
Date 1045 Sunday April 1, 2001
Approximately 100 meters in the Pacific Ocean 65 Nautical
Miles west of Hainan Island

    "Conn, SONAR splashes in the water bearing

approximately 245 degrees creating a line going north

towards Hainan Island." The captain acknowledged the

splashes.

J8II Fighter People's Republic of China
Date 1100 Sunday April 1, 2001
Approximately 2000 meters altitude above the Pacific Ocean
30 Nautical Miles south of Hainan Island

    Lt Yang, who was recently promoted to flight lead,

formed up on the left side of the ailing P3. He and the

other members of his flight had to deploy flaps to maintain

the slow speed needed to escort the P3 to Hainan Island. Lt

Yang instructed the flight there were to be no hostile

actions towards their new guests…unless they witnessed any

hostile actions from the P3.

U.S. Aircraft P3 Sub Hunter ISR gathering aircraft Call
Sign Black Sheep 3
Date 1045 Sunday April 1, 2001
20,000 feet over the Pacific Ocean 52 miles south east from
Hainan Island, China

Airman Zyes looked at the Ensign after looking at

their Chicom escort. The airman asked the Ensign if he

thought shooting at the J8II out the over wing hatch with

his side arm would be a good idea. The Ensign said, "Hell

no! Put that away. His buddies would just shoot us out of

the sky and anyway we have to land at their base. WHAT? WE

ARE LANDING AT THEIR BASE?" "Yep better than sitting in a

dingy for ten hours and getting picked up by them later

anyway.

The P3 landed on runway 33 and skidded to a stop on

the long runway on Hainan Island met by a large welcoming

party that wasn't very welcoming. There were jeep type

vehicles and 50-60 ground troops with firearms trained at

the wreck which was the P3.

Lt Commander Smithey shut down the engines and addressed the crew, "You all did a great job today safeguarding America and keeping her secrets safe. This Chinese debacle is now a State Department problem. Remember Name Rank and Serial number only to the best of your ability." At that the crew entry door was forcibly opened and in rushed a Chinese tide of PLA soldiers.

U.S.S. Memphis Submarine
Date 1100 Sunday April 1, 2001
Approximately 200 feet deep in the Pacific Ocean 210
Nautical Miles NE of Hainan Island

"Conn SONAR new contact bearing 240 degrees very quiet. Running diagnostics now on classification." Captain Jeffery made his way to the SONAR room on the Memphis, which consisted of a small two crew station off the main bridge. Seaman Halls was on duty at the moment. He was a good SONAR Operator and was fairly seasoned. "What do you have Halls?" asked the Captain. "I think we have a new sound track. The computer spit out four possible classes of Russian subs and now it seems to think it is a biological." "I am recording. If it is Russian or Chinese our intel suggests that it will be able to detect our sound in fifteen to thirty minutes at our current speed and depth.

Looks like she is coming directly at us." Captain Jeffery

returned to the bridge and ordered the Helmsmen, "Slow

ahead make your depth 500 feet 5 degree down bubble and

come left to heading 270 degrees." "Aye Aye captain" and

the Helmsman read back the order.

Kuái 2 People's Republic of China Submarine
Date 1100 Sunday April 1, 2001
Approximately 100 meters in the Pacific Ocean 70 Nautical
Miles South of Hainan Island

Captain Zhang went to SONAR to see if there were any

interesting contacts that he could perceive to be testing

from China. Seaman Lang sitting on console reported three

contacts far away to the south west that were all surface

vessels and he thought they were probably fishing trollers

from Vietnam. The Paracel islands south east from Hainan

islands were blocking the sound coming from Kuái 1 and the

U.S.S Memphis. Neither Kuái 1 nor Kuái 2 knew of the

international incidence with the P3 that happened earlier

in the day.

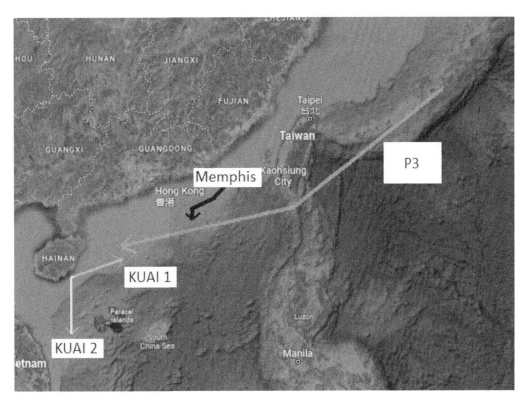

Figure 8 Map of Action for Chinese submarines and P3

33.

Date 1000 Monday April 16, 2001
654 CLSS Annex, Tinker AFB, OK

Lt Craken entered "The vault" for the weekly staff meeting with the CLSS. Steve was attached to the CLSS but did not fall within its rank structure. That is what the Air Force calls "Matrixed". He was invited to the weekly staff meetings since he became the Chief of Tinker's ABDR Program. As a lieutenant he commanded eighteen engineers housed in the various weapon system offices that made up the Oklahoma City Air Logistics Center. He was responsible for training and equipping the engineers to deploy to fix America's jets in war time. Since he didn't have the means to equip the engineers the 654 CLSS took on that function

for the engineer flight for all of their equipment needed

such as cold weather gear and chemical warfare gear.

Lt Craken was thinking *I hate meetings* until SSgt

Baletti walked in with donuts.

Lts love themselves some donuts. In fact if someone

from any office had donuts and email went out to all the

Lts telling them where they could find the free donuts.

Steve had a lieutenant friend Rob that responded to one

such email asking if they were evil donuts. "What are evil

donuts?" Steve asked in an email. No reply.

The 654 CLSS commander entered The Vault, Lt Colonel

Jones and started the meeting. First up was training.

Everyone had made their schedule training except Lt Lisa

Swage who was an engineer. Lt Craken answered that he would

talk to the Lt and make sure she knew what happens when we

miss training. It looks bad on the commander for the CLSS.

The commander was a good commander and gave the Lt some

grace. Typically the ABDR Engineer Chief was at least a

Captain or Major so having this responsibility as a LT

spoke volumes for Lt Craken's ability to lead.

The Commander asked next about readiness levels and

other mundane squadron info that Steve didn't really need

to pay attention so he grabbed another donut. About an hour later the meeting was over and everyone stood to leave. The commander asked the Lt to stay back for one more issue. He agreed.

The CLSS team lead for C/KC-135s also stayed for the impromptu meeting and the commander asked the last person out to close the thick vault door because they were going to discuss some classified information. The female TSgt closed the door and the classified meeting started.

Lt Colonel Jones asked if anyone was following the forced landing of the Navy P3 on the news. Both men agreed they had seen it. The commander then asked if the 135 team could go repair it and fly it out. The pentagon had asked the question to both the CLSS and to Lockheed, the manufacturer of the P3. Lockheed stated they could cut it up, crate it and put it on a slow boat from China for the low low price of thirty million.

The Lt spoke first and said, "well we can do it for a lot cheaper." The team lead MSgt Bans agreed. The Commander said, "That is what I thought. I will up channel this information. As of now do not tell anyone about the possibility of the 654 CLSS going to China to return the

P3." "Yes Sir", came both responses and the meeting was

over.

Date 1000 Monday April 23, 2001
654 CLSS Annex, Tinker AFB, OK

During the staff meeting Lt Swage came up again

missing small arms training. Lt Craken instructed SSgt

Baletti to remove her from anymore training and that she

was no longer a part of ABDR. Come to find out when Steve

talked to her about missing the training she told him she

was a pacifist and only went through ROTC in college to pay

for her education. The meeting went on for an unusually

long time and there were no donuts. The Commander asked The

Lt to stay back again and the C/KC-135 team lead. He had a

feeling he was going to China.

The commander asked Steve if he could support going to

China in a week to fix the P3. A CLSS team is going later

this week to Davis Monthan AFB-the bone yard, to pick out a

P-3 wing tip off a decommissioned aircraft in the desert.

34.

Date 1045 Monday April 30, 2001
Meilan Airport International, Northern Hainan Island

Lt Craken stepped off the plane marking 25 hours in the air on four different airplanes starting in OKC to SEATAC and then going military contract flight to Tokyo, Japan. The leg from SEATAC to Japan was rough. First, since it was a military contract flight the airline did not care about anything the passengers had to say or feel because they had a contract with the DOD and any complaints they may have had got lost in the Pentagon Shuffle. The seat rows were spaced so close that Steve hit his knees just sitting normally and he was a short dude. Since he was just put on this flight last minute of course he got the middle seat.

Every time he started to doze off during the 10 hour flight the lady behind him would get up for a mad dash to the lavatory and use the back of Steve's chair to help her get out of her seat. The results were Steve's seat became a spring board every 15-20 minutes. No sleep had there. Steve was fairly new in his career and knew everyone on the plane was military so this air sick crazy woman could be a high ranking officer so he bit his tongue and didn't say anything although he did fantasize punching her in the face.

From Tokyo the team was booked on a plane to Manila, Philippines and then on to Meilan Airport Hainan Island China. The team was able to get hotel rooms at Tokyo but the rest of the weekend was sleeping in the terminals. It was a good thing Steve was young because this was some seriously weird jet lag. Basically the team flew into the future based on time zones. It was a day ahead of what his biological clock was telling him.

When the plane taxied to the gate the copilot walked back to where Lt Craken and the team were sitting and instructed them to wait until all the children of China and invited guests of the People's Republic of China had de-boarded the plane. SSgt McMasters said, "WTF to that?" Lt Craken said, "Just wait for the welcoming committee".

The team had been briefed before leaving Tinker that they would not receive a warm welcome and that it could be potentially threatening. Lt Craken did have the phone number to the state department in case they were not given what the needed or were being harassed. They were also warned that there would most likely be listening devices and/or cameras where they were staying. Lt Craken had an idea for that.

Everyone but the team de-boarded. Lt Craken stood up to get his carry-on bag as did the rest of the team. When they did a detachment of Chinese soldiers marched on the plane with their battle rifles at the ready. The leader yelled at the team to sit down in some really bad "Engrich". Lt Craken raised his arms like he didn't understand. Steve knew the Chinese would try to intimidate the team and he wasn't buying it. For two things Steve was short…for American standards but he was a full 4 inches taller than these little jackasses and sure Steve worked out but he wasn't winning any Mr. Olympiad competitions but contrast him with these scrawny 13 year old boy looking dudes and there was not much intimidation taken by the team.

Everyone made it off the plane without an international incident and they were met by Mr. Po, some Chinese equivalent of the U.S. state department weenie who mock chastised the soldiers for their "rude welcome" of their guests. He informed the team that housing arrangements were made for the team close to where the wreckage of the American aggressive airplane which was helped in their time of emergency. Everyone on the team was thinking the same thing but Lt Craken gave the Chinese dude a sarcastic look and slowly said, "Right", with raised eyebrows, which apparently was lost on everyone but the team.

The team was escorted to a school bus and was taken a couple of blocks to their quarters in an industrial part of the city. Their quarters were really bad and appeared to be scheduled for demolition. The Lt called the team to the parking lot and told them they were going down town to the Guilin Long Sheng Hot Spring Spa Hotel which was going for $65 a night U.S. Dollars which for China was like a $300 a night hotel. There were no per diem tables for Hainan

Island so the Lt was like let's do it and ask for forgiveness later. It was located a short distance away close to the S82 that the airport was on. The Lt knew there would not be listening devices in those hotel rooms for a day or two. The group hired a couple van taxis, grabbed their stuff, and headed to the sweet hotel. Steve noticed that away from the airport you could mistake this place for Hawaii. It was beautiful.

The plan was to take 24 hours to get used to the crazy travel to get here and to get the ABDR trailer to the sight and begin work on Wednesday May 2.

Date 0800 Wednesday May 2, 2001
North East Dirt Ramp Meilan Airport International, Northern Hainan Island

The team arrived at the dirt ramp where they had towed the busted P3 call sign Black Sheep 3. She was rough! The team gained access to the P3. From what the aircrew did and then what the Chinese did to gain U.S. secrets, the inside was a big mess. TSgt Kaufman said, "Man it looks like someone went crazy with the crash ax…like in the movie "*The Shining*!" The ABDR trailer made it but all off the doors were open despite the team pad-locking the doors before it shipped State-Side. Lt Craken asked MSgt Bans to inventory the trailer and see if anything was missing. The MSgt said he would and the Lt went to inspect the spare wingtip from Davis Monthan AFB. The rest of the team started assessing the damage.

About the time the team started assessing the damage the Chinese minder from their state department came up to the Lt. He said, "Please excuse, why you not stayed in the very nice accommodations my government provide?" MSgt Bans saw the circus start and headed over to back up the Lt. He knew the Lt but just wanted to make sure he wasn't taking any crap from this politician. Lt Craken said matter-of-factly, "That flea infested condemned crack house? Are you kidding me"? Mr. Po said, "It is very rude in my country to not accept a very generous gift like this that my country give." Lt Craken said, "Were not staying there. Talk to my state department. Now I have work to do so I can leave your country." Lt Craken walked off leaving the politician with his mouth hanging open.

It's always a slow start when traveling to fix a jet. You have to make sure you have all your tools and thanks to the Chinese customs some of the heavy tools were "held up for inspection" so the Lt made his second call to the State

Department about getting their K-12 saw and generator released or bought back off the black market…whatever. His first call was about their crappy accommodations and the State Department said to stay in what was provided and to see if they could get something worked out. The Lt was like "Screw that we're staying at the spa," and he put it on his travel card.

Next on the To Do List was getting the damage report from the assessor/hydraulics troop, SSgt McMasters. The good news is that it did not appear that there would be much engineering. All they needed to do was provide a ferry flight to Okinawa, Japan. Most of the repairs were in the Navy Structures manual and fairly decently documented…for the Navy. The wingtip replacement would take some time but it seemed fairly straight forward-take off the bolts that held on the wreckage and then lift up the tip and install. The problem was getting a crane because this wingtip was five foot by three and fairly heavy so he would need to ask his minder for a crane.

About then the minder, Mr. Po came driving up in his wanna-be black limo. Steve ignored him for a few minutes and then an army jeep came driving up with the same scrawny kids on the plane. MSgt Bans said, "I'll handle them. You keep working on the wing tip." Lt Craken said, "No, I've

got this. I think we are really making some breakthroughs

in our relationship." The MSgt said, "Ok, Lt."

The Lt walked to Mr. Po and said, "Afternoon". Mr. Po

asked, "What is 'afternoon?'" the Lt said, "It's a greeting

usually preceded by 'Good' but 'good' would be back home

chilling by the pool." MSgt Bans heard the whole

conversation and chuckled a little. Mr. Po looked over at

the scrawny soldier and nodded telling Steve everything he

needed to know. Unfortunately, Steve knew he couldn't

strike first so he had to let the soldier hit him and he

did…with the butt of his rifle. Steve took a hit to his gut

but he knew the next play was to knock him out with the

rifle butt to his head. He prepared for that and blocked

his head with his arms just in time. The strike knocked

Steve backward but Steve went with it and did a backwards

roll to his feet. Mr. Po tried to restrain the soldier but

he wasn't having it and ran at Steve which is exactly what

Steve wanted. Steve had studied American Karate in high

school and taught himself some boxing so when Scrawny ran

at him he held his ground and at the last minute performed

a Tai otoshi, a judo move that the person moves his leg

over to block the attacker's legs and then grabs them and

turns away from the attacker affectively throwing them

forward without the attacker being able to get their legs

under their weight.

The rest of the ABDR Team came running over. Scrawny's

buddies aimed at the rest of the team and one fired warning

shots into the dirt.

Steve then disarmed Scrawny and gut checked him with

his own firearm and kicked his head knocking him out.

Before Steve did anything else he threw the rifle down in

the dirt so he hoped, he wouldn't get shot by Scrawny's

friends.

Scrawny's Friends looked at Mr. Po, who clearly

didn't know what to do. One of the solders moved up and

took Steve into custody. MSgt Bans started yelling at Mr.

Po to release the Lt immediately but they all packed up and

left. MSgt Bans picked up his phone and called the State

Department.

Date 1252 Wednesday May 2, 2001
Local Police Station, Northern Hainan Island

Lt Craken was forcibly removed from the jeep and taken

inside what appeared to be a local police station. An old

guy in police uniform came out yelling at Scrawny for what

Steve perceived as meaning *I was trouble for this cop*. The

Lt was placed in an interrogation room. Steve was nervous

but not scared…yet. He also was planning his escape. He had

to pick the right time and he made the decision if they

transported him to location B then he would have to act or

he would never be seen or heard from again. So for now he

would chill and conserve energy. Fifteen minutes later Mr.

Po walked through the door with the Police Chief and said

he was free to go. Steve gave Mr. Po a sarcastic "Thanks,"

walked out the door and hailed a cab back to the airport.

The team was glad to see the Lt back. They decided to

knock off a little early and go get a beer at a bar close

to the Spa.

The rest of the time the team was in China was

uneventful. There were a few engineering decisions but no

full blown designed repairs for the Lt to design so he

helped the team by bucking rivets and designing a lifting

system making use of an old communications tower the

airport was no longer using. It was close to where the P3

was parked and needed to get the wing tip up to the wing to

bolt it on. Mr. Po was not seen or heard from so really the

team was treated with mild neglect. After they hung the new

wingtip the P3 started looking airworthy. The Lt called his

contact at the Navy and had them send out the aircrew and

ground crew because they were almost finished.

Date 1630 Saturday May 5, 2001
North East Dirt Ramp Meilan Airport International, Northern
Hainan Island

     The P3 Aircrew and ground crew arrived on sight. The

Lt briefed the new aircraft commander on flight

restrictions and he went to check it out. The pilot told

the ABDR team that they had one seat for someone direct to

Okinawa. Lt Craken looked at MSgt Bans and said, Someone on

your team can have it." MSgt Bans said, "No way. You earned

that seat from what you did to Scrawny. If you hadn't done

that those Chicomms would have harassed us the entire time

we were here." The Lt said, "OK."

Date 1000 Sunday May 6, 2001
North East Dirt Ramp Meilan Airport International, Northern
Hanan Island

     The P3 coughed and sputtered but got all engines up

and running. The commander contacted the tower for takeoff

and the tower came back and said they could not have clearance for one hour. The commander decided to leave the engines running and would take on hot refueling before they taxied out. The pilot said they probably had to scramble fighters to "Escort" us out of their airspace.

An hour and fifteen minutes later the tower gave the taxi clearance and finally takeoff clearance. The P3 flew again. There were in fact four fighters that came up on the wingtips after take-off. The commander ignored them and climbed to 36,000 feet and soon after the fighters would leave due to burning most of their fuel to get to 36,000 feet.

Part Three:

09/11/2001

35.

Tuesday September 11, 2001
Oklahoma City, Ok Tinker AFB

    1st Lt Steve Craken was promoted to first Lieutenant

two months ago and along with the new costume jewelry he

received, which was a silver bar instead of a gold one. He

also received a hefty pay raise. No more living paycheck to

paycheck. A few of his Lt buddies noticed his promotion but

no one else did. It is customary in the military to have a

promotion party but I guess in a sea of civilians it really

doesn't matter. So when his buddy Jacob got promoted two

days ago he made sure to send the whole SPO an email so
they would know.

　　　Today started like any other day. Lt Craken woke up
at 0600, took a shower, put on his BDUs and grabbed some
breakfast. He left his rental house in the Village and
made the commute to Tinker AFB in about thirty minutes on
the Broadway extension to I-40. When he first received his
assignment for Oklahoma City there was a mile-wide tornado
that went through the city so he called around and found an
apartment in Midwest City, a suburb of Oklahoma City. It
was close to the base. He learned to never live close to
the base at this first duty assignment because his
apartment was across the street from section 8 housing AKA
the ghetto. So he moved to the Village which is up north
of Tinker AFB. Steve went through the Douglas gate with no
issues and then parked in overflow parking because the
normal parking lot was narrow and it was possible to get
completely blocked in the parking space by a large vehicle
opposite of where you parked since he had a Dodge Ram Quad
cab. Steve entered the building and went upstairs. The
"office" was a huge room with ten rows of 100 cubicles each
but they weren't real cubicles. They were basically two
walls in an "L" shape and you shared where the third wall

should be with a coworker and behind you was the aisle - no

privacy. His cubicle was almost against the far wall. He

rolled in around 7:30 which is early for most of the

civilian workers. In the 135 SPO there are a handful of

military including the commander. Steve turned his ancient

computer on and opened email. He learned to let it churn

through its downloads before attempting to do anything on

it. Once email was downloaded he went through it and

decided on what he was going to do today. There were a few

interesting structural request emails for repair from the

field.

#1 from MacDill AFB in Tampa Bay. In a pre-flight

inspection the ground crew found corrosion and popped spot

welds on a fuselage stringer for tail number 54-1643. They

included a picture of the stringer but it was zoomed way in

and impossible to tell where it was. My disposition for

that one is probably remove and replace. Steve was

thinking, *"I'll have to see if I can get another trip to*

Florida out of it."

#2 from Eielson AFB Alaska. Tail number 56-1419 was

down for a local phase inspection and one of the inspectors

found a 1.3 inch crack on the 820 bulkhead. The number

"820" means 820 inches aft of the front datum point, which
for the 135 is 14 inches in front of the nose radome.

#3 from Peas AFB New Hampshire. They wanting support
for a flap jackscrew. That falls under systems so I got to
forward that to my buddy Lt Adam Smith. Lt Smith went to
the Air Force Academy and was commissioned 6 months after
Steve so Steve helped him with in-processing. He is a
fairly cool dude, as much as one can be that went to the
Air Force Academy. Steve pranked Adam a couple of weeks
ago. Adam left his computer unlocked and went to the
bathroom. Steve sent Adam's wife an email that looked like
it was from him. His wife worked over in Personnel and was
an officer also. In the email Steve wrote "Take out the
garbage? That's woman's work!" Adam came by Steve's cube
a couple hours later and maybe after some marriage
counseling with the Chaplain. He sent the whole email
thread to Steve and she was pissed.

It was now about 9:45 in what seemed like a normal day
when someone came by and was talking to a coworker next to
my cube. There are no private conversations in the OC-ALC.
She said a plane had crashed into the World Trade Center.
So Steve started Googling it and sure enough it had. Steve
had started working on his private pilot's license a few

months before and he was wondering if New York had bad

weather that caused the plane to hit the tower but it was

Visual Flight Rules or VFR with Visual Meteorological

Conditions or VMC prevalent, which meant that weather was

not a factor in the crash.

Steve remembered his trip to NY with his church

friends for the 1999-2000 New Year's event at Times Square.

He knew how packed NYC was with buildings and to have two

major airports there makes it that much worse.

So basically all worked stopped at the ALC and

everyone was glued to radios and TVs down in the cafeteria.

At 10:05 a second jet flew into the other tower and they

all knew it was an attack. The base commander shut down

all the gates. No one got in or out of Tinker AFB. Steve

called Col Craken, who was a chaplain in the Army to see if

he had heard the news and he had not so Steve's Dad said he

had to go check in with his command.

When Steve was a senior in high school his Dad

deployed to Dessert Shield so he knew he might be deploying

again after the attacks.

When Steve was on the internet he discovered the

Federal Aviation Administration (FAA) had closed U.S.

Airspace. *"What does that even mean?"* he was thinking. Then he read that no planes were allowed to take off from U.S.

airports and all planes were diverted to Canada or Mexico

that were still in the air. Lt Craken tried to get work

done but really he was just glued to the internet and

radio. Other reports came in later about the Pentagon being

hit by a plane and even a third plane crash in Pennsylvania

in a field.

Later in the day the base commander opened the base

back up for people to leave. Steve thought to himself,

"what does tomorrow look like and knew the world was

changed forever."

Wednesday 0730 September 12, 2001
Oklahoma City, Del City Tinker AFB South Gate

Steve had a fairly normal morning on the way to work.

Today he was meeting with some enlisted members of the CLSS

at the ABDR Lab at 0800. He drove to the South Gate instead

of his usual North Gate but as soon as he turned off SW29

Street he saw the traffic was backed up from the gate two

miles and knew they were searching every vehicle going onto

the base. Steve looked down at his coffee and decided to

stop drinking it because he was going to be in this line

for a while. Little did he know it would be three hours to

get on base. There were people making a run for the woods

and he seriously considered making a run himself. When he

finally got up to the gate the Security Forces (Military

Police - SF) checked under the Steve's truck and he opened

the hood while a bomb sniffing dog checked his truck for

explosives. He was cleared and on base he went. His first

mission the day after 9/11 was to find a restroom.

Thursday 1000 September 13, 2001
Tinker AFB Cube Farm

    Today Lt Craken was thinking about the world and the

uncertainty. He made a note to go shopping for things he

would need because he was fairly certain he would have a

deployment due to the events of 9/11. He opened email and

saw an email marked "IMPORTANT" from the ABDR Headquarters,

Hazel Macintyre, the point contact for all ABDR engineers.

The email:

      Lt Craken contact me to schedule a STU III classified
phone call at your earliest convenience.

My STU III information:
211A-23
223 994-2345

Sincerely,

Hazel Macintyre
ABDR Engineers Liaison
DSN 445-2365
Wright Patterson AFB

      Steve went to see if Gloria was in her office because

she worked with special purpose C/KC-135s and dealt with a

large amount of classified information. She might know how

to make a STU III call.

      Gloria was in her cube. She had a full cube to herself

because of the information she handled. Steve asked if she

knew where a STU III was and how to use it because he

needed to make a classified call. She pointed to the corner

of her cube to what looked like a broom closet and opened

the door to reveal a sound proof booth with a STU III.

Gloria said, "You need the STU III identifier and the phone

number of the STU III you are trying to call but you also

have to send your identifier and phone number to the person

you want to talk to." Steve said, "I have the information

for the other phone but I need this STU III info to send to

them." Gloria said, '"That information is written on this

STU III." and showed him.

Steve wrote the information down and went to his cube

and wrote an email back to Hazel saying he could call on

the STU III in thirty minutes. Hazel wrote back immediately

agreeing to the time.

Thirty minutes later Steve was working the STU III. He

inserted the key that Gloria had given him and all the

information to call Hazel. After six minutes of fiddling

with the expensive phone he heard the garbled tone of the

far STU III handshaking with the local STU III. Then he

heard Hazel answer.

"Lt Craken, the information I am about to tell you is

classified Secret/No Foreign. You may tell the ABDR

engineers so they have time to prepare but they cannot tell

anyone else. This is your formal WARNING ORDER for the

following Engineers; B-1, B-52 and C/KC-135. They need to

get all of their mobility bags issued and their engineer's

kit. They need to be ready to deploy in 72 hours and they

should not travel over an hour away from their home base.

If they are tasked to deploy I will contact you again

through the STU III with your EXECUTE ORDER. Do you

understand everything I have told you today?" "I do, thank you," said Steve in his professional voice.

Lt Craken walked over to his desk and sat down to process all the information he just received. He decided to walk over to the B-1 Program Office and the B-52 Program office to talk to the two ABDR Engineers in those SPOs. The C/KC-135 SPO was between the two bomber SPO offices. The Engineer for B-1 was Lt Julie Bear. Lt Bear was a tough girl who was 5'9" 210 pounds of muscle. You didn't mess with her. Lt Carly Roach was a 95 pound petit smart engineer that had a degree from MIT. Steve decided to find Lt Roach first. He headed over but her cube was empty. He found her supervisor and asked where she was. The supervisor told Steve she was giving a tour at the saw tooth building. The saw tooth building was built during World War II and had a roof that looked like a saw blade. They made them like that so sunlight would come through the windows at the top of the roof. Lighting back during that time was poor so the designers added extra sunlight coming in.

Lt Craken left the B-52 SPO and went to the B-1 SPO. He found Lt Bear at her desk and said, "You need to go tell your boss you are doing ABDR stuff today. Then you need to go to the CLSS and get your deployment bags. All three of them - Cold Weather, Chemical Warfare and Mobility Bag along with the Engineer's Kit. Then you need to go home and pack. You are on a WARNING ORDER. The next order is the EXECUTE ORDER expected in 72 hours. You need to stay within one hour travel time of Tinker AFB. Do you understand what I just briefed you?" She said, "Yes, I understand."

Steve left his building and drove over to the Saw Tooth building to look for Lt Roach. He walked around a few minutes and saw a group of people walking around the TF-33 jet engines that power the great B-52. The B-52 uses eight of these engines! He saw Lt Roach giving some generic spiel about the engines and then she saw Lt Craken and asked one of her coworkers to continue the tour. Steve relayed all the information to her that he did to Lt Bear. Steve gave her a ride to her car back over at Building 3001 where the program offices are so she could drive over to the CLSS Annex to get her bags and then go home to pack.

Lt Craken was the C/KC-135 ABDR engineer and the Chief of all the ABDR Engineers at Tinker. It would be an easy call who would go for C/KC-135 except Lt Smith was also an ABDR - 135 engineer.

He had a tough decision to make. He chose himself to deploy if they received the EXECUTE ORDER for C/KC-135 team to go. Adam could deploy on the second rotation if they still needed ABDR engineers.

Steve left the base and went to the annex to get his mobility bags, stopped at an ATM machine to get out as much cash as he could and went home to pack his gear in case he received the classified call.

Monday 0900 September 17, 2001
Cube Farm, Tinker AFB

Steve had a good weekend hanging with his friends. They all asked if he was deploying and he said he could not tell them. Unfortunately, it was Monday and that meant he had to be at work. He checked his email earlier and nothing significant until 0915 when he saw an email from Hazel again marked "IMPORTANT". So he opened it and responded to her request for the STU III call.

The Lt talked to Hazel on the STU III and she said, "You now have an "EXECUTE ORDER" for all three weapon systems I alerted you for. Your B-52 Engineer is on Chalk One and the B-1B and C/KC-135 Engineers are on Chalk Two.

Both Chalks are leaving from Tinker AFB. Chalk One is with the Third Herd or your resident Combat Comm and Chalk Two is with the Security Forces Unit. Who are the engineers going?" Lt Craken gave her the names and they ended the call. He called Lt Bear and Lt Roach and filled them in. He looked at the schedule and determined both Chalks left on Wednesday September 20.

36.

Wednesday September 20, 2001
Oklahoma City, Ok Tinker AFB Mobility Center/Base
Operations

Hurry up and wait…The motto of militaries around the
world. All three Lts sat on their mobility bags or mob bags
waiting for their planes to load with their respective
Chalks. There was some readiness colonial that made Lt
Roach the Chalk commander for Chalk one. He asked her if
she had ever commanded before and she replied, "No". The

colonial said, "Well don't screw this up. Here are the folders for the troops on your chalk." For Chalk Two there was some Security Forces Captain deploying so neither of the other ABDR Engineers had the hassle of leading the Chalk. Lt Craken was sitting on his bag when SSgt Baletti came running through the door with a big hard-sided case. He looked frazzled. He gathered all the CLSS troops to him and started issuing firearms. The enlisted troops received M-16s and M-4s and he brought the officers the Beretta M-9 handgun. Lt Craken liked to shoot guns. He had qualified on the M-9 and M-16. He shot expert on the M-16 at basic training. Out of the four times for 9mm training he never shot expert with it and it bothered him a little. It was a cool rainy day at Tinker AFB so everyone had their Gortex rain jacket on or out inside the building. When SSgt Baletti handed the Lts their side arms and magazines he said they would get ammunition down range. Steve asked him, "Where are the holsters?" and he got a serious deer in the headlight look. The Lt said, "Don't worry about it and placed the M-9 in the side pocket of his Gortex and told him he would scrounge one up down range.

The call came over the PA system that Chalk One was ready for boarding. Lts Bear and Craken did not envy Lt

Roach herding cats out to the C-141 Starlifter cargo plane.

Soon after they cleared out and the call came for Chalk Two

to load onto the C-17 Globe Master III for which Steve was

stoked to not fly on the 40 year old C-141.

Before both Chalks left, the Deployment Colonial told

Chalk Two they were going to Diego Garcia in the middle of

the Indian Ocean. He would not say where Chalk One was

headed. Steve thought to himself they were probably going

to a "Stan" (Pakistan, Uzbekistan, Tajikistan, etc.) secret

base.

Thursday September 21, 2001
Diego Garcia, Middle of freaking nowhere, Indian Ocean

Both Lt Craken and Bear grabbed all their bags and

headed off the plane. There was some Major Director of Base

Operations greeting people coming off the plane and

ordering them to surrender their laptops. Lt Craken looked

at Lt Bear and said, "No". The Major stopped the Lts

because really there were not that many officers coming to

the island. Most troops were enlisted. He asked if they had

laptops and both said, "Yes". He said the base commander

has ordered everyone new to the island to give their

laptops to Base ops. Lt Craken said, "I can't do that sir.
I am an aircraft battle damage repair engineer and I have
to have my laptop to do my job." The Major's face started
turning beat red. Apparently no one had told him no today.
Lt Craken spoke again before the Major could really freak
out, "I have an in-brief with the commander tomorrow and we
can sort it out then". The Major looked at Lt Bear and
said, "What about you?" She replied, "Same". Lt Craken
started walking off as did Lt Bear. They were tired and
dirty from the 27 hour flight to the island and both had
their laptops on their shoulders.

Friday September 22, 2001
Diego Garcia, Middle of freaking nowhere, Indian Ocean

     Lt Craken told the team he would be late to PT that
morning because he was going by the armory to see if he
could get a couple of holsters for his M9 Beretta and Lt
Bears M9 so they didn't have to keep them in their rucks or
front pockets. He walked in as a new maintenance team was
arriving from the States. There was one Captain, their
maintenance Officer, with the team that was requesting an
M16 along with his M9 because he had an intel brief a

couple days prior that said there may be some sabotage from locals or from Chinese that had small boats. Their team was issued all M16s. Lt Craken stepped up to the window and asked for holsters and an M16. The armorer gave him the holsters and told the Lt that the Captain grabbed the last M16. Steve asked him if he had any M4s or anything else. The armorer said, "Some jar head Mo-tard left a M110 Semiautomatic Sniper System or SASS. You can check that out." Lt Craken replied, "That's a 0.308 right?" The armorer was nodding his head and was thinking this is no dumb Lt. "I'll take it", said Steve. "I'll need some extra ammunition also to sight it in." The armorer gave Steve a whole ammo can of M80 ball ammunition and the drag bag full of all kinds of goodies including a suppressor. Steve was thinking to himself, "This rifle is going home with me."

37.

Friday 0935 October 5, 2001
Afghanistan, 50 miles North of Kabul "Tango Ridge"

B-1B Lancer tail number 83-0066 "Ole Puss" was making

bombing runs fifty miles north of Kabul with coordinates

from ODA 595 (U.S. Special Forces in country working with

General Dostum) as part of Task Force Dagger. The B-1B was

providing close air support to the U.S. Special Forces plus

the indigenous forces they were imbedded with. Extreme

weather moved in and the snow and hail from the fierce

storm limited the visibility of the laser to the GBU-10 Paveway II, 2000lb Bombs. The crew decided to decrease altitude to gain the ability to drop bombs on the enemy, the Taliban.

Ole Puss, Call sign Extremis 5, dropped down to 15,000 feet well above the mountain peaks in the area. Their intel brief that day had the only threat to the aircraft as the extreme weather they were now below but well within the B-1s safe flight envelope. The aircraft commander instructed the copilot to monitor wing and engine ice buildup and to turn on the anti-icing systems.

The B-1B Lancer is an all-weather day/night bomber so it was designed with robust anti-icing systems. To keep ice from building up on the wings, the engines can supply hot bleed air to the leading edge of the wing heating it and preventing ice formation. The same is true for the engine itself. The nose dome and inlet guide vanes in the front of the engine are heated preventing ice from building up. The danger of ice for the wings is two-fold. Number one, ice adds weight and decreases range of the aircraft. Number two and the biggest danger is ice can change the shape of the wing and cause it to stall well within the normal flight envelope and the plane could fall out of the sky. The

danger for the engine is for ice to break off and get
ingested into the engine causing Foreign Object Damage or
F.O.D. Ice can tear apart a jet engine.

The crew contacted the Joint Forces Air Controller
(JFAC) on the ground with the ODA 595 Special Forces team
and gave them the green light to illuminate targets for
their bomb run.

ODA 595 illuminated a tank with dismounted troops in
close proximity. The Taliban had no idea that the ODA Team
was above them on a ridge several hundred yards to the
south. Extremis 5 started in on the bomb run and was ten
seconds from dropping six GBU-10, but the Defensive
Electronics Warfare Officer (EWO) in the left back seat of
the B-1 called code "Wave Off, Wave Off, Wave Off! There's
another friendly jet in the area beneath us."

Call Sign "Saint One" an Air Force National Guard F-
15E from Naval Air Station Joint Reserve Base New Orleans-
an Air National Guard base, heard the radio chatter from
ODA 595 to Extremis 5 and wanted to draw some enemy blood
before their rotation was over. The crew of the F-15E was
not in contact with ODA 595 but had the target in sight and
their GBU-10 could see the target illuminated. Saint One
swooped in and dropped two GBU-10 and egressed after

watching the secondary explosions. They were Bingo for fuel (low on fuel) and had to leave to hit up a tanker close to Uzbekistan and back to the secret base, Karshi Khanabad.

ODA 595 reported the two GBU-10 hit aft of the approaching column with minor damage to a couple of supply trucks.

Extremis 5 rolled back onto the target and released six GBU-10 bombs. Thirty seconds later the bombs exploded with one hitting the tank and two others hitting the dismounted troops.

ODA was providing battle damage report to Extremis 5 when the B-1 had to pull up to miss a mountain peak. When they flew over the backside of the peak the aircraft commander banked the aircraft to the right to make another bombing run if needed. The copilot yelled, "I have light flashes from the ground at the two o'clock position". A half second later the plane bucked with a loud Bang. The Commander pushed the four throttles into full afterburner and climbed like a bat out of hell!

The Copilot said, "I think we were hit by anti-aircraft fire." "Check all gauges and warning lights," exclaimed the commander. The master warning light was not on and so far no other caution lights were on. The copilot did notice the number two reserve tank was reading zero.

When aircraft are designed the engineers usually put reserve tanks in or on the wingtips because this helps the wing structure and skin with fatigue. In flight the wings are being bent upward with the aerodynamic load but the weight of the fuel in the reserve tanks helps to mitigate the bending load and fatigue in the wing structure.

"ODA 595, Extremis 5, we have been hit by suspected anti-aircraft fire from the backside of the ridge you are on about 2 klicks south. We are bugging out. Extremis 2 should be on station or another asset in five minutes. Extremis 5 out."

"Extremis 5, ODA 595 copies. Good Luck!"

Extremis 5 decided to bug out and head for their temporary home of Diego Garcia which was a four hour flight south in the middle of nowhere Indian Ocean. Extremis 5 linked up with their tanker and started taking on fuel. The Copilot had to dig deep in the emergency procedures to configure the right wing fuel pumps and valves to close off the reserve tank and prevent further leaking of fuel. The Commander asked the Offensive EWO to see if he could see any damage on the wing and he said, "Yes Sir, I see a hole close to the right wing tip."

Friday 1042 October 5, 2001
Afghanistan, 50 miles North of Kabul "Tango Ridge" 2 klicks south of ODA 595s original position

   ODA 595 team Bravo rode their horses south to where the anti-aircraft position was reported by Extremis 5. They stayed on the south side of the ridge until they got close and then dismounted their horses and low crawled to the top and looked over the ridge. Under camo netting was an old Russian ZU-23 twin 23 mm cannon that appeared to get lucky hitting Extremis 5. There were tents set up next to the gun…way too close for comfort and it appeared one tent had been set on fire from the recent activity of firing the gun. An ODA 595 operator quietly asked his team lead, "How in the hell did they get that up here?" "Horses?" he whispered. "Let's smoke these guys and blow their gun up!" said the team leader. The detachment leader spread his four men south along the ridge using hand signals. When everyone was in position he gave the signal to fire. Team Bravo all had suppressed M-4 so there was not much sound coming from the rifles. All of the Zhu-23 crew died instantly. Hearing

a lack of commotion the Zhu-23 commander came out of his

tent to see why his men were not getting ready to fire on

more aircraft and he was shot two times in the head by the

team lead and another man.

Team Bravo moved into the camp and looked for any

intel they could find. They took pictures of the dead

Taliban to send back to their Intel Unit to see if they

killed any high Value Targets (HVT). Most of the team

grabbed AK-47s and ammo because they didn't know how long

they were going to be there without a supply drop and hey

free gun and cool trophy. Then the team leader had the

demolitions guy rig up the ZHU-23 with C-4. He also asked

him to rig up the ZHU-23 ammo as well. He did and they

helped themselves to a couple of horses that the Taliban

would no longer need in paradise. They led the horses to

the other side of the ridge and packed up. 500 yards up the

trail Bravo Team lead contacted the main detachment of ODA

595 to instruct them of the fireworks going off in 1

minute. One minute later they blew the ZHU-23 and it was

spectacular.

38.

Friday 1430 October 5, 2001
Diego Garcia (British territory 2,500 Miles South of
Afghanistan)

"DG Tower, Extremis 5 sixty miles out declaring an
emergency." "Extremis 5, DG Tower, say fuel remaining and
nature of your emergency." "Fuel at 1.5 hours and we have
battle damage on our right wing possibly leaking fuel. We
request you roll fire response." "Copy Extremis 5. Fire
trucks rolling wind 170 at 8 knots cleared to land 13. If
possible taxi to southwest ramp area." Extremis 5 had a

normal landing and was able to taxi off the active runway to the southwest run up area where the ABDR team set up shop. There was no fire.

Figure 10 Extremis 5

Lt Craken heard the mayday call on the radio and met the plane with the team leader so he could get the initial response from the flight crew as to what happened so they could begin fixing the jet. Before the engines shut down the base commander for the U.S. Forces, in his blue and white pinto from the 1970s. Colonial Bolls was his name and he treated the ABDR team like a red headed step child. In the arrival brief that Lt Craken tried to give him he was very dismissive. He said, "Nothing can damage any of our aircraft. You are a waste of tax payer money!" *Well now he looks like he might be happy we are here,"* the Lt thought. Col Bolls jumped out of the Pinto and was screaming, "Can you fix this jet? I need it tomorrow for ops?" Lt Craken stepped up to him since he was addressing the MSgt and calmly said, "Sir we have to interview the flight crew and then determine what was damaged. I will let you know an estimated time to fix it but I don't think it will be on the schedule tomorrow." "I want a full brief for swing shift changeover at 1700". The Lt said, "Yes sir" and just like that the ABDR Team became the most important unit on this little steamy hot island. As soon as the commander left the Lt Said to MSgt Gleeson, "Rock Paper Scissors to see who has to go to the swing shift brief?" The MSgt said, "OK 1, 2, 3…" The Lt stopped him and said, "No I'll go. He's a jackass and who really cares if they see some Col chewing out a Lt. It's kind of expected. I'll leave you here to start the cleanup of damage. Let's go find out what happened from the pilots." Lt Craken gathered all the information from the flight crew and wrote it down on some engineer's paper that would become part of the AFTO Form 781 which is the aircraft's permanent maintenance record-a binder that is several inches thick. Steve was thinking how

nice it would be to have the design loads from Rockwell,
who manufactured the B-1B but he was a C/KC-135 engineer
and didn't even have the design loads for the 135 - trade
secrets. He would have to use the Material Ultimate Load
(MUL) method to repair this bird. If he had the design
loads he could use those to design a repair but the MUL
method has a higher safety factor than design loads. The
MUL method takes the loads that the original structure
could take without inelastic failure and returns the
structure to that higher margin than if he was just using
design loads.

After the flight crew left, the Lt and some of the
junior team members grabbed a maintenance stand and moved
it up to the right wing of the massive B-1B. Back at Tinker
if you wanted to be a wing walker you had to check out a
harness and lanyard but not here in the South Indian Ocean.
The whole team scrambled up onto the wing to start
assessing the damage with no harnesses.

Steve thought, *Wow a 23 mm shell can do some damage to
aluminum!* Then the Team lead said as much. MSgt Gleason had
the B-1B Battle damage T.O. the -39. He looked at the Lt
and said, "Sorry there aren't much canned repairs for this
area." The Lt just said, "Time to earn the big bucks." He

looked at the assessor and asked for his notes when he was finished with them and dropped down off the wing to get the engineers tool kit and calculator. Lt Craken also stopped by the trailer to see what supplies there were on it. Thankfully it was fully stocked before it left the States. The assessor was quick and brought the Lt his notes. Steve took all the information and went inside the fuel pit office, a makeshift office for him and the team. The building was the command center for fueling operations…probably during WWII but was forgotten since. The ABDR Team took it over and by that I mean they cut the pad lock off and threw out a whole bunch of old paint, drop cloths and scaffolds. No one complained so they used it as their makeshift offices. It had an AC unit that sort of worked. Instead of it being 110 inside it was a balmy 87 degrees F.

Steve looked at the damage notes and determined this repair would have four parts. The first would be the shotgun looking patterned holes on the underside of the wing. That was easy because he would just have the tech put on speed tape both inside and outside. Speed tape is awesome. Think duct tape but on steroids. If you took 10 sheets thickness of aluminum foil and added an industrial adhesive to one side and then rolled up 100 feet of in 4 inch width you would have speed tape.

Number two are the two stringers that were totally blown away. Stringers provide bending rigidity for the wing. In flight there is a lot of upward bending force on the wing due to the lift it generates. The opposite is true when the airplane is on the ground because of the fuel load all of the weight bends the wings downward. The stringers attach to the ribs which provide the shape of the wing or airfoil. Stringers run from the wing root all the way out to the tip of the wing. Ribs run parallel to the fuselage or in the wind stream direction. Spars are really beefy Stringers. Typically there is a front and rear spar. They can be "I" beams or box beams.

The stringers in this location at the wingtip have a cross section shape of a "Z".

The third repair component for this repair is the skin patch to make the wing aerodynamic again.

The fourth component is fasteners for the stringers and skin.

Lt Craken wrote up his repair and finished it before he had to go to the 1700 shift change meeting. He discussed it with MSgt Gleason and asked him what he thought about a time line. The Lt suggested 3-4 days if they could get some air carts and generators to run the tools and covering two shifts. The team leader agreed and the Lt started walking to make the 1700 brief.

39.

Friday 1700 October 5, 2001
Diego Garcia Wing Commander's Conference Room, Hangar 1

    Lt Craken had to jog the last mile in order to make the meeting on time. He had the good sense to take off his BDU shirt or it would have been soaked by the time he got there. He scrambled into the latrine and tried to make himself presentable. Lt Craken entered the room and looked for a seat in the back. Some crusty Major, probably the director of Operations, tried to kick Steve out of the meeting saying, "We don't need any coffee Lt!" Lt Craken looked him square in the eye and said, "It's too hot for coffee and I am briefing the commander". The Major huffed a

little but didn't seem to care. At that the commander came

in and Steve was fairly certain the commander was full of

himself because he made everyone stand at attention a

little longer than necessary.

Colonial Bolls looked at Lt Craken and asked, "Who the

hell are you and why are you here?" Before Steve could

answer, the commander said another snide remark, "You know

all Lts are worthless." Lt Craken looked at him and said,

"Well probably not this one, sir. I have the assessment for

your battle damaged B-1B." The Col said, "Well then let's

have it, **Lt**!"

"Yes Sir!" said Lt Craken. "The number 2 reserve tank

on the right wing was probably damaged by a 23mm shell. The

lower wing skin has minor damage like what would be seen if

it were shot with a shotgun. That damage is the easiest to

repair. The next damage is a bit trickier, two stringers

that were completely severed, and then a large skin patch

will have to be fabricated and installed over the wing."

The commander exclaimed, "I knew it, you and your team are

worthless! You're trying to make excuses as to why you

can't repair it!" Steve had a serious look on his face and

said, "No sir, we can repair it with materials we brought

in 5-6 days, if we get the necessary Aerospace Ground

Equipment or AGE."

"Lt, I've been in this Air Force for many years, some

I don't want to count. There is no way you can repair that

with some trailer you brought!" Steve calmly replied, "I

assure you we can and we will. It will be back on the

schedule by Tuesday." "You got until Sunday to get it done,

Lt", said the commander. "OK, well I'm going to go work on

it then," said the Lt and got up and left while the

commander made another "Lt" joke at Steve's expense. Steve

didn't care because he knew his team could fix the jet. He

got back to the jet and designed the repair.

Tuesday 1830 October 9, 2001
Diego Garcia Fuel Pit South end of the runway.

MSgt Gleason was inspecting the completed repair with

the Lt. when they heard a vehicle approaching. It was the

stupid looking pinto with the Wing Commander sitting in the

backseat and only a driver. He was using the crusty Major

as a chauffeur! The Lt looked at the MSgt and said, "Got

that done in the nick of time!" MSgt Gleason said, "For

sure. I'm glad you padded the repair time or we wouldn't

have had enough time." The col headed over to the maintenance stand and climbed up. Lt Craken said, "We just finished the repair sir. We need to finish up some aircraft records and the ops can have her back." The col looked at the repair and said, "Damn that's ugly." MSgt Gleason beat the Lt to it and said, "Yes sir but she is airworthy". The Lt chimed in and said the plane will get a permanent repair the next time she goes to depot."

Kuái 2 launched their mini sub that had been hitchhiking top of the sub since they left home base at Hainan Island on. The mini sub contained four Chinese commandos. They learned this trick from the North Koreans invading South Korea. The mini sub set a course due West or 270 degrees to intercept Diego Garcia. At two miles per hour they would make land fall at dusk.

Kuái 2 turned to a heading of East 090 degrees at all ahead full and 100 meters deep. The commander of Kuái 2 wanted to put some distance between them and the mini sub to limit detection of either asset.

Monday 1630 October 15, 2001
South Fuel Pit area…Make shift ABDR Headquarters Diego Garcia

Steve was hanging out on the beach by the end of the runway. He had just finished designing and installing a minor repair to a B-52 that had a maintenance stand blown into it by another B-52's jet wash while it was taking off.

The Lt gave the BDR Team the day off to go to the nice beach up north on the island where there was a real bar since it was somewhat slow. The B-52 "battle damage" came in while they were gone and it was minor, very similar to the damage they had back a Tinker at the ABDR lab.

There was a bar on the island but the base commander gave everyone a coupon for one beer a week. Steve had saved his and brought them to ABDR Headquarters DG and put them in the fridge and was enjoying a few. He was thinking he really was missing being back at home and his family and friends. He also was thinking about walking the two miles back to his quarters but he had to share those with three other officers and hadn't made the time to get to know them. Truthfully they were really Col Boll's cronies, so he was just going to chill here for the day, enjoy some solitude and if any more repairs came in he could start working those.

He thought to himself that he hadn't really had time to spend with God so he cracked open his Bible and started reading some of 1 Samuel 30. This chapter was talking about the Amalekites burning King David's home town of Ziklag and taking his family including two of his wives captive. What did David do? He prayed and asked God if he should pursue the Amalekites. God answered him with a yes. He didn't hesitate. He gathered some men and went after them. He rescued every single person taken captive and killed a

great many of the enemy. Steve was contemplating this
medieval and barbaric combat. He had always thought about
being infantry in the Army and becoming a sniper but he
loved aircraft more. He was tired so he spent some time in
prayer and thought about taking a nap.

Monday 2045 October 15, 2001
20 feet deep just north of Barton Point Diego Garcia

    Shang Shi (or Senior Sergeant) Ping was busy checking
gauges in the mini sub and keeping track of navigation.
They had just entered the lagoon of Diego Garcia from the
north and had a little time before night and before they
stormed the beaches. The mini sub was not known for any
creature comforts. It was hot, moist and rank. The
commandos had been in the mini sub for over 10 hours and
really didn't care if they lived or died but just needed
some fresh air. The team's mission was to kill military
personnel, especially pilots and, if possible, to sabotage
as many planes as they could. Their plan was to swim up the
beach on the lagoon side of the base. A team would stay in
the barracks area shooting anyone and everyone while the
other team would make its way to the flight line and blow

up any aircraft there. They off loaded Fire Team Two, code named Dragon, which consisted of Shang Ding Bing (Seaman) Wong and Seaman Pong Don. Fire Team Dragon exited the mini sub directly into the water at 30 feet deep north of the runway close to where lodging was located.

The mini sub containing Fire Team One, code named Tiger, continued to its final destination down to the south side of the runway (about two miles) on the Lagoon side of the island.

Monday 2115 October 15, 2001
100 miles due East of Diego Garcia

Kuái 2 slowed her speed and came up to 40 feet. Captain Zhang ordered the boat's Executive Officer to the fire control station. It was time to execute the orders. Seaman Phet Li was sitting on console when the Captain and Exec came to his station. Captain Zhang asked, "Do you have a firing solution for all 20 Fire storm missiles?" "Yes sir, they are ready to fire". The commander looked at the Exec and said, "Fire when ready". The Exec parroted the command and Seaman Phet Li started firing the missiles. Two by two the missiles were fired from their vertical launch tubes by encapsulating them in compressed air and then ejecting them out of the ocean by the buoyant force. Once free of the ocean, their booster rockets fired getting air

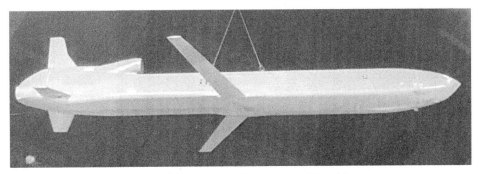

Figure 11 CHICOM Firestorm Missiles.

flowing through the air breathing jet engines. Once the engines were ignited the missiles would drop back down to 25 feet above the ocean to avoid detection. All the cruise missiles had been preprogrammed a specific flight path to all arrive at their individual targets at the same time. The Exec said, "Twenty minutes until impact". Captain Zhang said, "Make your depth 120 meters 20 degree down bubble and come right to heading 0. *Time to leave the crime scene,* thought the boat's Captain good luck commandos."

Monday 2115 October 15, 2001
Control tower Diego Garcia

Airman Lewis Johnson was a bright young man of 23 years of age. He was actually a Mississippi Air National Guardsman but drew a rotation to Diego Garcia as part of an obscure program between the State's guard units and the Navy. Somehow in high school he got the nick name "The Milk Man," but no one knew how that came about. All the enlisted called him Milk Man but the officers only knew him as Amn Johnson.

Amn Johnson was sitting on console as part of his base rotations. He would go to almost all of the enlisted career fields that were represented on the island learning their jobs. This rotation, two weeks' worth, was spent on the RADAR console. He was joking with his naval counterpart when his console alerted him to new contacts…a lot of new contacts east of the island. This was strange to him because all the previous contacts came from north of the island returning from Afghanistan. Since this was not his career field he immediately asked for help. The supervisor, Petty Officer Chavez, came over and asked him what he had. Amn Johnson said there were a bunch of new contacts east of the island. Petty Officer Chavez knew no one ever flies in from the east. Amn Johnson said, "I saw 10-15 new contacts

from the east." Petty Officer Chavez looked at the blank

screen. He knew the schedule for the day and most of the

base's birds were back from that day's missions and none

would take off until closer to dawn. Petty Officer said to

the seaman to the right, "Do we have any carrier based

aircraft coming for a visit today?" The Seaman replied,

"None scheduled." He looked back at the blank RADAR screen

and then to Amn Johnson with a scowl, "You messing with me

Airman? Cause I'll bust me an Air Force puke just like I

would a seaman!" "No Petty Officer Chavez, I swear I saw a

whole bunch of new contacts to the east and then they

disappeared before you came over." "Maybe it was aliens!"

exclaimed Chavez getting excited. "Seaman Beaumont, pull up

the last 15 minutes of contacts for Amn Johnsons' console."

"Yes Petty officer Chavez", said Seaman Beaumont. The

screen data played out on the 50 inch RCA that hung on the

wall. Everyone in the tower stopped what they were doing

and watched the blank RADAR scope for 7 minutes 32 seconds

while Amn Johnson started to think *did I really see*

contacts on the screen? I haven't been trained for this

crap. I'm a jet engine troop. Then in rapid secession 20

new contacts appeared about 100 miles east of Diego Garcia

all at zero airspeed accelerating vertically then

accelerating horizontally to all different azimuths and

then dropped down off the RADAR screen. Amn Johnson yelled,

"SEE!" Everyone else started to chatter as to what they had

just seen…until Petty Officer Chavez exclaimed, "SHUT UP!

SHUT UP! HIT THE WARBLE TONE ON THE CLAXTON AND GET TO YOUR

SHELTERS! VAMPIRES IN THE AIR! VAMPIRES IN THE AIR!" He

picked up the phone and called the base commander's direct

line. Col Bolls picked up the phone to hear Petty Officer

Chavez screaming, "Sir we have incoming missiles. We have

incoming"… and about that time the wavering tone Claxton

sounded.

Monday 2145 October 15, 2001
Diego Garcia

Fire Team Tiger parked the mini sub on the bottom of

the lagoon under the fuel dock, set the timer on the

explosives for 45 minutes and exited the sub into the

water. This was going to be a one way mission, but if by

some luck they could get back to the sub, then Petty

Officer Ping could stop the explosives and they could use

the mini sub to escape, but that was a long shot.

41.

Monday 2150 October 15, 2001
Diego Garcia

    Lt Craken woke from a dead sleep involuntarily kicking

his foot up in the air from the 40 year old recliner he

"borrowed" from the joint Officer/Enlisted Club on the

island a few weeks ago. "Where in the Heck am I", he asked

no one out loud. It took about 45 seconds for him to

determine where in the world he was. Oh right I'm deployed

to a little rock in the Indian Ocean. "What in the heck was

that noise…the Claxton with wavering tone went off…WE ARE

UNDER ATTACK!!!" He grabbed his battle rattle (bullet proof plate carrier with 9mm Berretta, extra magazines for the pistol, magazines for the SASS sniper rifle and Individual First Aid Kit (IFAK) and put it on. Then he realized he had to put his boots on and then helmet. He grabbed the SASS rifle and pulled the "T" handle to chamber a round. The SASS rifle system is based on the M16 but chambered in the larger caliber of 7.62 X 51 mm NATO. Typically military members don't walk around with a round in the chamber because there is no return spring on the firing pin. What that means is that if the rifle were dropped the pin could strike the cartridge and fire the round without pulling the trigger, but that would take a great deal of force because the cartridge primers are much harder than pistol primers but it could happen.

He looked out the window and it was dark except for the runway lights and the lights across the runway on the ramp where there were six B-1Bs parked. Then the lights went out. He thought to himself, *This has got to be a drill. Who would or could attack this island? It reminded him of his U.S. History class taught by a coach at Starkville High. The lesson was on "Isolationism". Isolationism where the U.S. stayed out of world conflicts because there were no enemies within striking distance at the time.*

I better treat it like a real attack. Since he was miles away from his assigned bunker or any bunker, he went to the back of the ABDR Office and climbed down into the holding tank for the fuel from de-fueled aircraft. The tank was big, dry, below ground and hadn't been used in at least a decade or maybe even since WWII.

He was hanging out in this old fuel tank feeling stupid for being there when he remembered a story that a ROTC cadet had told him back when he was at summer camp: It was at the height of hostilities during Desert Shield/Dessert Storm. Saddam Hussein was in the habit of lobbing scud missiles at bases in an attempt to actually inflict casualties on the Coalition Forces. The Scud missile was not really accurate. Fire it in the direction of a target and you may hit it. At the time it was believed that Iraq could put chemical weapons on scud missiles. Well it was 23:15 one night and the Claxton went off, *much like right now*, thought the Lt. Everyone was quartering in tents at that FOB. The Iraqi military launched a handful of scuds at a Coalition FOB. Everyone quickly ran to their bunkers and put on their chemical (Chem) warfare gear. Chem warfare gear consists of pants and a shirt that contains powdered activated charcoal. The charcoal would absorb any chemicals (or biological, nuclear material) that go through the fabric keeping the wearer free of the chemical. The gear also consisted of rubber boots and gloves a gas mask with a hood attached. Wearing chemical warfare gear is a chore! It's hot no matter what the outside temperature is.

So everyone (officer, enlisted, men and women) is sitting on the floor in this bunker with nothing to do but wait and hope there are no impacts from the missiles. Well about that time Lt Charles Wesson, a logistician (moves around beans and bullets) comes running into the shelter buck naked carrying his mobility bag "C" or Chemical bag with all his chem warfare gear in it. Immediately everyone turns and stares at the spectacle. His feet are wet from recently being in the shower and he slips on the floor and sprawls all over the place with his flip flops flying through the air. It looks like some weird oversees peep show. The Lt doesn't even seem to notice anyone and starts putting on all his chem gear starting with his mask and hood. His friend and co-logistician Lt Shelly O'Brian took out her camera to document life at the FOB…and he didn't even notice. There were no impacts at the FOB that night because a Patriot missile battery did its job and brought

down all of the scud missiles fired that night in close

proximity to any coalition forces. Well, there were some

positive impacts, on the moral of the females inside the

bunker that night.

The small earthquake brought Lt Craken back to the

present but it wasn't an earthquake. Thu…Thud…Thud…

Thud…BANG BANG!! It was an impact followed by another

stronger one. He counted 12 in all impacts. He waited five

minutes to be sure the attack was over. He slowly started

coming up from the tank clearing the area with the SASS.

Unfortunately, he was never issued night vision

goggles because hey they were on an island hundreds of

miles from any OPFOR (Opposing Forces). What he could see

after the attack in the low light was two B-1B bombers

completely destroyed and on fire. The remaining four had

sustained some level of damage probably due to the close

nature of the explosion. All the lights were still off.

That is going to be a lot of work if I make it through

this attack, he thought. *I need to get on the radio and see*

if I can help. The Lt went over to the desk and found the

radio. It was dead. He opened the drawer and found a spare

battery. He swapped it out and then the chatter on the

radio was crazy. He knew better than to get on and clog it up so he just listened to what was going on. Battle damage reports were coming in from all over the island. The tank farm that stored jet fuel JP-8 and diesel was hit with at least two big tanks engulfed in flames, which he could see from two miles away out the window! The control tower was gone but a temporary tower was being set up on the north side of the island. He heard this information from the radio. Reports were coming in from the Security Forces (SF) they were engaging OPFOR on the ground. The fire department was split between the ramp with aircraft on fire and the tank farm. Losing anymore aircraft would seriously degrade the Air Forces' ability to wage war in the Afghan Theater. Another report from SF said they saw OPFOR heading toward the ramp. The Lt thought, *If they make it to the ramp they will engage the unarmed firemen and we'll lose more troops and planes.* The Lt exclaimed out loud to himself, "Not on my watch".

Lt Craken hastily ran out the door and was trying to remember where any cover and concealment was, but on the ramp in runway environment there isn't much. Concealment is being hidden from the enemy while cover is behind something that bullets can't go through. Ideally you want both. He thought about making a sniper's hide out of a 3 foot Bravo taxi way sign, but there was no egress route if he were detected and engaged by OPFOR so he would become pinned down…or worse. Also he didn't know if it were cover as well as concealment. So that was out. He remembered several trees southwest at the end of the ramp and decided that would make a decent hide. So he took off running for the trees while listening to the cacophony of attack aftermath.

The tree group turned out to be a half mile away. The Lt made it in three and a half minutes with carrying all the gear he had. The tree clump was a great sniper's hide with a forest to the east and the sea behind and to the west with great sight lines especially of the ramp and the burning aircraft. If he were attacked he could retreat to the ocean and swim around the forest and evade through there just like a SEAL.

He noticed on his run over to the hide that there must have been a missile attack because some of the taxi way was cratered. One missile had barely tagged the edge of the runway.

Lt Craken set up the SASS and wondered how far away the non-burning aircraft were because that would make a great target for the OPFOR. He thought, *Maybe there is a*

rangefinder in the bag. He looked and found one and a
spotting scope. That's great otherwise, he would have to
use the runway markings to estimate the distance. He was
working on his private pilot's license so he knew there
were signs that told pilots how many feet of runway
distance they had left, in thousand feet intervals, as well
as runway lines that gave distance information also.

   He pulled out his notepad that he always kept in his
BDU pocket and started making a range card for the ramp. He
got "lucky" because the aircraft that were on fire were the
two furthest from his hide at 718 yards away. He could see
the fire department with their big fire engine dousing
those two in foam. The closest aircraft that was not
damaged was only at 166 yards away. He sighted the SASS in
at 200 yards so the closet bird would not need any
adjustment, only some Kentucky windage with the elevation
or really an adjustment down by ¼ Minute of Angle (MOA)
with the lines inside the scope. MOA are a measure of
accuracy for a rifle and shooter. Typically you need 1 MOA
or better for a sniper rifle. Steve had been shooting all
his life, because his Dad and grandfather were hunters.
Shooting long distance was new for Steve because he didn't
have a large caliber rifle at home, just some shotguns and

a .22, but he knew the fundamentals and picked it up quickly. A 1 MOA Accurate rifle is fairly easy to explain. At 100 yards the rifle can shoot within a one inch circle at the target. At 200 yards it becomes a 2 inch circle and keeps growing with every 100 yards added. So if your rifle can shoot 1 MOA accuracy, then, theoretically, you can shoot a life size target in the kill zone at 600 yards because you aim a center of mass and if the bullet can hit 6 inches in any direction that would be a kill. Steve understood MOA but had to learn the military system of accuracy which is Milliradians or MRAD because that is the type of scope that was on the SASS.

Bird number two was 417 yards away so that would need some come-ups to put the cross hairs on target at 400 yards. The SASS that Lt Craken had was sitting at ¼ MOA accuracy so he knew the gun could outshoot his skill. The NATO 7.62 X 51mm round he was using had a drop of 20.9 inches at 400 yards.

The scope on the SASS is a First Focal Plane, which means the reticle size and target will increase/decrease in the same proportion when the magnification is adjusted. That is important because it allows the shooter to quickly asses the range of the target. This scope is a 5-25X50.

What that means is the magnification of the scope starts at
5 times magnification and can be adjusted up to 25 times.
The "50" is the tube diameter in mm which is important
because the bigger the tube the more light allowed into the
scope. The shooter could adjust the scope in 1/10 MRAD per
click or adjustment of the turrets. There were two turrets
on the scope for elevation and windage.

Steve left the scope at the zero in case he had any
OPFOR near the close bird.

The Lt's hide was next to a road and he heard a HUMVEE
approaching and gunfire. The lead HUMVEE was shooting at
the chasing HUMVEE and Lt Craken had no idea who was good
and who was OPFOR so he maintained cover and watched. The
chasing HUMVEE became disabled with flat tires and a
leaking radiator while the lead HUMVEE made its way to the
second bird from the Lt's hide. Lt Craken quickly dialed in
seven clicks up on the scope and started observing the
actions of the HUMVEE and its' occupants. It stopped and a
soldier jumped out. He was short and stocky and in a
different uniform than anyone else the Lt had seen on
base…*Has to be OPFOR*, he thought.

The surviving member of the Chicom Team Tiger opened
fire on the fire truck closest to him and it quickly sped

away. Lt Craken forgot to account for any wind but there

was a slight breeze in his face so no correction needed. He

lined up the soldier in the scope with the cross hairs

directly on center of mass and then raised the rifle so the

internal scope lines were three half ticks below the center

cross hairs effectively raising the point of impact 1.5

MRADs (Steve could have adjusted the scope with the

elevation turret but by using the internal lines of the

scope he could change shots quickly based on various

engagement distances). The Lt took in a full breath let out

about half of it and held. Then he gently squeezed the

trigger while holding on the OPFOR getting something from

the HUMVEE. The SASS went off surprising Steve, which is

what you want with no anticipation that can move the shot

off target. HIT! But looks like it was a little low and hit

him in the leg. Steve quickly mentally added a half MRAD up

inside the scope and took aim again. The OPFOR had no idea

where the shot came from so he went about his last task on

this Earth of crippling one more aircraft. He had a satchel

charge and was ready to throw it under the engines of the

B-1B. Steve fired his second shot. The second team member

of Dragon dropped to the ground in a heap of meat. He was

dead before he hit the ground by a 180 grain bullet to the

throat (grains are a unit of weight in the firearms industry. Typically powder and bullet weight are measured in grains. A grain is 1/7000 of a pound). Three seconds later a big explosion rocked the HUMVEE but did not cause any damage to the aircraft.

Seaman Li was missing in Action (MIA). He possibly was blown out to the lagoon when he placed a satchel charge on a huge fuel tank.

Lt Craken got on the radio, "Firebat 3, Engineer 5, Threat neutralized on the ramp. Resume firefighting and be aware there may be casualties south of your position on the road in a HUMVEE." "Engineer 5, Firebat 3, Did you make that shot and did you blow the HUMVEE up?" "Firebat 3, Engineer 5, I did shoot the OPFOR but he had a satchel charge trying to take out the B1, Over." "Copy", Said Firebat 3. Steve headed off to see if he could render first aid to the SF HUMVEE on the road.

Tuesday 0030 October 17, 2001
Diego Garcia

Col Bolls was losing his freaking mind! The base was
attacked, and he had no warning or intel and there were
casualties. He ordered everyone to stop what they were
doing except the firemen who had been battling the JP-8
tank fires and go to the base gym for accountability. So
far there had been 12 casualties. Lt Craken met the ABDR
team at their makeshift assembly area which was the fuel
pit where they usually worked. Everyone was accounted
for…except TSgt Susan McCormick, a damage assessor and
electrical troop. Susan was a 31 year old former Miss Utah
winner. She was beautiful! Her father was an Air Force
veteran and had passed away from cancer when she was in her
freshmen year of college at the University of Utah, the
Utes. He had a huge impact on her life and she wanted to
honor him so she marched down to the Joint Forces
Recruiting Office and enlisted in the Air Force.

Susan was at the base club getting some food when the
attack hit. The rest of the team was on the beach, where
there were no military targets. The base club was next to
billeting so the ABDR team deduced that billeting was the
target. When Susan heard the attack she hunkered down until

it was over and then went outside to see if she could help

anyone hurt in billeting when Fire Team Dragon sprang up

from cover and mowed her down with their SKS rifles.

MSgt Gleeson and TSgt Kaufman ran up from the beach to

see if they could help their fellow troops out and render

self-aid and buddy care (first aid). They came around the

corner of billeting just as Fire Team Dragon was engaging

one of their own. TSgt Kaufman yelled, "NOOOO!" selected

full auto fire on his M16, lined up one of the Chicoms in

his iron sights and swept the general area where Seaman

Wong was standing. TSgt Kaufman scored five hits center of

mass and that was the end of the Chinese commando. MSgt

Gleeson selected semi-auto on his M16 and put two rounds

center of mass and one in the head of Petty Officer First

class Ping, the Chinese commando team lead. That marked the

end of the weak Chicom invasion.

42.

Tuesday 0615 October 17, 2001
Diego Garcia

It was a long night. The ABDR Team assembled and

started prioritizing assignments even before they were

giving orders from Col Bolls. When Col Bolls arrived at the

flight line it looked like someone had kicked an ant pile

with all the ABDR troops crawling over aircraft.

Lt Bear talked to Lt Craken and they decided they

would both act as assessors until all aircraft had been

assessed. The good news is almost all of the B-52s were

either at other bases or returning from missions during the attack so the one or two that might be damaged were automatically on the back burner. One cruise missile found the far B-1B and it was a direct hit. That damage spread to the B-1B next to it but the fire was contained before it could have damaged the third or fourth B-1B. The "Star of Abilene," tail number 83-0065 that rolled off the assembly line before "Ole Puss," was a complete loss. She was reduced to a burnt pile of aluminum and a bull dozer pushed her sad carcass off into the grass off the ramp. "Ole Puss" was saved from any further battle damage because she was over by the ABDR make-shift office away from harm.

Col Bolls stormed up to the Lts who were both on tail number 84-0072, "Spanky" up on the right wing surveying damage. "Both of you Lts get your butts down here now!", exclaimed the Colonel. They both moved with a sense of urgency and got down the maintenance stand and expected a butt chewing from the commander. Col Bolls, said, "I need these airframes fixed ASAP and there will be a base funeral for all personnel lost during the attack at 1100 at the base theater. Also one of you is re-deploying to a classified location early next week. I don't care which."

As soon as Col Bolls left, the lieutenants looked at each other. Lt Craken said, "Rock, paper, scissors?" Julie, said, "No you are the more experienced structural engineer and I have a better rapport with the Commander here." "That is for sure!...The rapport thing. That dude does not like me. You are just as good an engineer and leader as I am! Ok, I'll let the team know about redeployment and I'll see how many repairs I can design before I have to leave."

Lt Craken talked to the team and let them know that half of them would be re-deploying to a classified location. He talked to MSgt Gleeson about "Spanky's" right wing lower skin and the burn damage and shrapnel damage. He said, "Let's just re-skin the wing in panels as long as the sheets on the trailer are big enough." "That won't work Lt, because we used that thick sheet for another repair. "We don't have one big enough." "Ok", said the Lt, "I'll go check sheet metal on base and go shopping there after the funeral. Maybe we get lucky". "Sounds good, said the MSgt.

Tuesday 1100 October 17, 2001
Diego Garcia

The funeral went for a couple of hours as the warriors honored their dead. Each squadron was responsible for making the soldier tribute of their fallen, the boots, rifle and helmet of the soldier or airman. The 654 CLSS had also made a polished sheet metal crown for TSgt Susan McCormick, AKA Miss Utah, and placed it on top of her Kevlar helmet.

Tuesday 1315 October 17, 2001
Diego Garcia

Lt Craken left the funeral and went to the sheet metal shop on base. He asked the supervisor, "Do you have any large sheets of aluminum in 2024-T3 or 7075-T6?" The supervisor said, "There are some large sheets of metal out in the storage shed but I don't know what flavor. I don't think they are steel." The Lt went out back and found several large sheets with "2024-T0" stamped on them. *That could work*, He thought. *Now how to heat treat them?* The "T-0" meant there was no heat treat so the aluminum was soft. Surely there is a boiler or something Civil Engineering (CE) has that could heat treat these sheets. He drove the

"Borrowed" HUMVEE (from the crusty major) over to CE and found what appeared to be a knowledgeable TSgt.

TSgt Helms looked dirty. He must have been fixing something that was hit during the attack. Lt Craken came up to him and said, "I have a problem and need your help." The TSgt answered tersely, "No, no you don't. I have 25 problems of my own." Lt Craken asked him, "Hear me out and I'll give you two beer tickets." TSgt Helms looked like he could use a couple of beers. TSgt Helms invited the Lt into a cool break room. The LT said, "I'm an aircraft battle damage repair engineer and I need to heat treat some aluminum sheets so I can re-skin a B-1's wing." "What temperature and for how long?" asked the TSgt. "About 1000 degrees F for an hour and then 350 degrees for a couple hours after." The TSgt said, "I don't have anything that gets that hot…but the fire department might." Tell SSgt Combs at the fire department that I sent you over." I'll do that and here are your beer tickets", said the Lt.

Tuesday 1345 October 17, 2001
Diego Garcia

Steve was enjoying this HUMVEE and he had one more stop to make before returning it when an SP drove up behind him and pulled him over with the blue lights which really looked out of place on a sand colored HUMVEE. Two eighteen year old SPs both at the rank of Airman exited their vehicle, one male and one female. When Steve saw the female SP he knew he was in for a hard time and he had better approach this with his hat in hand. The female SP, who looked like a model walked up to the driver's side of Steve's "borrowed" HUMVEE. She said sternly, **"Get out of the HUMVEE without your weapon and place your hands on the hood."** Lt Craken slowly drew his M9 Beretta and placed it in the passenger seat next to him while the two SPs watched him like a hawk. Steve replied, "I will comply with that but you need to recognize the appropriate military customs and courtesies." Steve heard the male SP whistle like the LT had defied the female SP AMN Sikes. AMN Sikes forcefully replied, **"I think you are confusing your rank with my authority"**, Lt Craken then replied while getting out of the HUMVEE, "No I am not going against your catch phrase or authority and like I said I am complying but you still owe me a greeting and a salute. I don't care if you respect me

or not but you better respect the men and women that gave

their life for their countries that wore the rank I now

wear." She contemplated what Steve said and then said,

"Good afternoon **SIR**" like a very sarcastic 16 year old, and

saluted but did not wait for Steve to return the salute.

Then she said, "Let's get down to business. Where did you

get this HUMVEE?" Steve replied, "From headquarters Parking

Lot." "Did you have permission to take it?" "Yes it was

unlocked." About that time another SP HUMVEE came rolling

up to the current scene that Steve was stuck in, with two

AMN SPs that had a Lieutenant with his hands on the hood.

Col Bolls jumped out along with the crusty Major and a Lt

Col that Steve had never seen before, but he wore the SP

beret, so he figured he was head of the SPs, the cop shop.

The trio of senior officers walked over to the HUMVEE

that Steve "borrowed". The crusty Major began to inhale

like he was about to unleash a typhoon of obscenities on

Steve but Col Bolls stopped him and asked "What are you

doing with this HUMVEE **LT**?" Lt Craken didn't face him and

kept his hands on the hood, "Sir the B-1 that was next to

the one that received the direct hit needs to be re-

skinned." Col Bolls stopped him and commanded that he face

them. Steve looked at AMN Sikes and asked, "Is that OK with you and your authority?" She began to speak when the SP Lt Col yelled at both of the SPs and told them to get back to the armory and clean the turned in weapons. They scurried back to their HUMVEE, which was blocked by the cols HUMVEE so they sped off through a field trying to get out of there as fast as they could. After that spectacle Steve started to explain what he was doing. "Sir we can fix 84-0072 right lower wing skin with sheet metal we found at the sheet metal shop but it needs to be heat treated. I was on my way to the fire house to see if they had some training aids or burners in the training house or training aircraft that we could use to heat treat. That way we can get that bird back in the fight instead of use it for parts…the can bird (cannibalize aircraft).

Col Bolls said, "Major Huffines, drive the Lt to the fire department and then back to the Battle damage repair area. Lt, you better get my bird fixed, and also you be packed and ready to fly out by Monday." The Major saluted smartly and got in the HUMVEE. Steve also saluted with a "Yes sir" and walked around to the passenger side, re-holstered his M9 and got in.

SSgt Combs did not like seeing the Lt and Major come in looking for him. He was sure he hadn't done anything wrong but you never know when something can come up and bite you in the butt. SSgt Combs had a strange look on his face when the Lt started talking instead of the useless Major. Soon he figured out he wasn't in trouble and started problem solving. "Yes I think we can heat something up to 1000 for an hour and keep it fairly consistent and then the 350 afterword. It will eat up a bunch of propane but, hey, this is war time. Bring the metal to the training facility north of the runway and we'll take care of it." Steve thanked the SSgt and told him the Major would get him some extra beer coupons and the SSgt thanked the Major.

The Major dropped Steve off at the south end of the runway without saying a word to him. Steve couldn't resist and said, "Thanks for the HUMVEE," in a way where you wouldn't know if he were sincere or poking the badger, so the Major just said, "You're welcome" and peeled out.

43.

Tuesday 1500 October 17, 2001
Pier 13 Long Beach Ca.

Army Major Splint oversaw the loading of three of his units onto the Pensiatti container ship bound for Guam. The ship was scheduled to arrive in two and a half days. She was part of the contracted fast ships the Army likes because they can sail around the world and get lots of tanks and stuff where they are needed. The Air Force can fly cargo but the airplanes always weight-out meaning they can only take a finite amount of weight and still fly. Not so with ships. The ships usually cube-out meaning they run out of hold and deck space before there is an issue with weight. So the ships can carry lots of 60 ton tanks where a C-5 can take only one M1A1 tank!

Major Splint was due to arrive with his best crews in a day and a half. When he landed on the island he would scout it out and find the best location for their deployment.

44.

Saturday 1723 October 20, 2001
Approximately 61 meters deep in the Pacific Ocean
400 miles North West Guam

Kuái 1 was cruising all ahead full in a straight line
From Taiwan towards the U.S. Territory of Guam. Since her
sister sub had a great success hitting the joint
U.S./British base Diego Garcia, the Politburo in Beijing
had become embolden and ordered Kuái 1 to do the same
attack on Guam because they knew the Air Forces B2s were
deployed to the island in the wake of the massive attack on
Diego Garcia for Force Projection by the U.S. The Politburo
was very afraid of the invisible Spirits.

The only difference in the attack would be no commandoes going in along with the missile strike because Kuái 1 did not have a team or a mini sub when they left for the sea trails that turned operational in the wake of the terrorist attack on their enemies, the U.S.

The mission order was to get just within striking distance, launch missiles and turn southwest toward the Philippines then head home to Hainan Island. The problem with the order is the effective range of the missiles. The Fire storm missiles could hit a Circular Error Probability (CEP) of ten feet from 100 miles as demonstrated by the Kuái 2 attack but their maximum range was 500 miles. The Cep at 500 miles was more like five miles. The island of Guam is 30 miles by 5 miles oriented roughly north to south. Captain Wing Ping was headed directly toward the narrow part of the island. Anderson Air Base is on the North East side of the island so if he fired his missiles at 500 miles he would statistically hit mostly water. He would have to be within 200-300 miles to be able to hit the B2 bunkers with any good statistical average.

U.S.S. Tucson Submarine
Saturday 1723 October 20
Approximately 200 feet deep in the Pacific Ocean
1500 meters behind Qing - NATO Designation (Kuái 1)

   The U.S.S. Tucson was ordered to replace the U.S.S.

Memphis two weeks after they started tracking Kuái 1. The

Tucson lay at the bottom of the ocean in a relatively

shallow area at 500 feet. The shallow area was a

continuation south from the Japanese island chain that

Okinawa belongs to. The route was along the projected line

of Kuái 1 since it was determined she was headed to Guam to

try to take out the B2s. The U.S.S. Memphis detected the

U.S.S. Tucson on the bottom with a magnetometer. The Tucson

was only about 200 meters off from Kuái 1's track! The

Captain of the Memphis thought they really wanted to hit

those B2s because Kuái 1 was not doing any defensive

maneuvers and didn't even have their towed SONAR array

extended-probably to make her as fast as possible. When the

Memphis passed the Tucson by 500 meters she quietly

descended and slowed. At the same time the Tucson made

herself neutrally buoyant and started out very slow and

then matched speed with Kuái 1.

Kuái 1 Submarine
Saturday 2300 October 20
Approximately 61 meters deep in the Pacific Ocean
350 miles North West Guam

Captain Wing Ping liked the thought of striking the

U.S. with his new sub. He did not like the orders on how to

strike the U.S. Base. The Politburo always had centralized

control but they also insisted on centralized execution. He

thought back to his days of professional military

education. He had studied the U.S.'s involvement in

Vietnam. The U.S. Government picked the individual targets

for the Navy and Air Force to hit instead of giving

Operational Control or OPCON to the fighting units to

decide what targets were important and which were not. He

wished for OPCON. Without it he was putting the sub in

danger. Basically, he was betting on stealth and not using

his sensors to see if any foes had detected them. He

understood the time table and the consequences of missing

the B2s that were forward deployed but he owed it to his

crew to use all defensive measures and tactics to keep a

whole picture of the battle space.

Saturday 2300 October 20
Anderson Air Base Guam

Air Force Major General Deacon was the acting Pacific Forces Commander (PACAF). He didn't like using the billion dollar airplane as bait for the Chicoms to attack, so they started 24 hour operations where only one B2 was allowed on the ground at a time. Guam had plenty of KC-135s to support refueling operations and the Air Force would only commit four B-2s to the island and that was four too many at 20% of the entire fleet! The Air Force insisted there be P3 sub hunters deployed to the island as well in case the sub swap didn't work and they lost Kuái 1. The Navy sent one and as soon as they had confirmation from the U.S.S. Memphis of the successful swap they pulled it out of Guam…just in case. Black Sheep 3 was fresh in their minds.

U.S.S. Tucson Submarine
Sunday 0530 October 21
Approximately 75 feet deep in the Pacific Ocean
150 miles North West Guam

The orders had been given to the Tucson before she intercepted Kuái 1 to observe and remain undetected. If the Tucson witnessed a hostile act then she could return fire but could not initiate an attack (including a preemptive

attack). The Tucson was well underway after the Diego

Garcia attack and part of her operating orders was to stay

radio quiet. A TACAMO jet was put up and broadcast an EAM

to the Tucson but the message was not received…for whatever

reason (the fog of war). The message read:

-----------------China is strongly believed to be the

instigator of a cruise missile attack on Diego Garcia. If

you witness an offensive posture by the enemy submarine you

are following, initiate an immediate preemptive strike,

especially if that sub is within 200 miles of Guam.

END OF MESSAGE---

Kuái 1 Submarine
Sunday 0540 October 21
Approximately 10 meters deep in the Pacific Ocean
150 miles North West Guam

    Captain Wing Ping ordered battle stations. He also

ordered a defensive maneuver to make sure they were not

being followed and that no enemy had detected them. The

helmsman turned 90 degrees to the left and deployed the

towed SONAR Array.

U.S.S. Tucson Submarine
Sunday 0540 October 21
Approximately 75 feet deep in the Pacific Ocean
150 miles North West Guam

"CONN SONAR contact designated Alpha one is turning

left and deploying their towed SONAR array!" Capt Miles

said, "SONAR CONN acknowledged, ALL Quick Quiet stop 10

degree down bubble." Capt Miles was hoping he got the jump

on the Chicoms by cutting his motor (along with any sound)

and diving below the thermal cline making them invisible or

completely quiet to the enemy SONAR. In a whisper Capt

Miles said, "Battle Stations."

Kuái 1 Submarine
Sunday 0600 October 21
Approximately 10 meters deep in the Pacific Ocean
150 miles North West Guam

Capt Wing Ping was satisfied no one knew they were

there and ordered the sub back to the original heading.

"Prepare for launch of the Dragon Fire."

U.S.S. Tucson Submarine
Sunday 0600 October 21
Approximately 75 feet deep in the Pacific Ocean
150 miles North West Guam

"CONN SONAR, looks like the Chicoms are going back to
their original heading and it does not appear they have
been alerted to our presence." "Copy", said the Captain.

Kuái 1 Submarine
Sunday 0610 October 21
Approximately 10 meters deep in the Pacific Ocean
150 miles North West Guam

"Seaman Xi, do you have a solution for all twenty
Dragon Fire missiles?" Seaman (Shang Ding Bing) Xi, the
Fire control operator, responded, "Yes sir." "FIRE". Twenty
doors on the top side of the sub opened and she began the
firing sequence of one missile forward, fired closely,
followed by one missile aft and so on.

U.S.S. Tucson Submarine
Sunday 0610 October 21
Approximately 75 feet deep in the Pacific Ocean
150 miles North West Guam

    "CONN SONAR, THE ALPHA CONTACT HAS OPENED ALL HER

MISSILE FIRING TOP OUTER DOORS AND APPEARS TO BE GETTING

READY TO FIRE MISSILES!" "HOW CLOSE ARE WE TO GUAM?" asked

the captain. "150 miles sir". "CONN SONAR ONE MISSILE

AWAY", TWO MISSILES AWAY". "FIRE CONTROL LAUNCH TWO

TORPEDOES AT TARGET DESIGNATED ALPHA ONE!" "Aye Aye sir",

said Seaman Trainor. Seaman Trainor had fired a torpedo

once in training but this was all together something else.

Peoples' lives hung into balance both enemies and

friendlies. He could feel everyone's eyes on him that were

in the immediate area. He did his job swiftly and hit the

fire button twice to send two Mark 54 torpedoes out the

forward tubes.

Kuái 1 Submarine
Sunday 0613 October 21
Approximately 10 meters deep in the Pacific Ocean
150 miles North West Guam

    "CONN SONAR TORPEDOES IN THE WATER 1000 METERS BEHIND

US CLOSING FAST!" The XO looked at the captain and asked,

"Do we stop firing missiles for evasive maneuvers?" Captain Wing Ping thought for what felt like an eternity which was only 3 seconds. "No continue firing dragons, launch countermeasures and turn 20 degrees right." The countermeasures were not likely to fool the torpedoes without fast evasive maneuvers. If he had given the order to stop firing the missiles and he could have performed an emergency dive below the thermal cline they might have had a chance. "How much time to impact?" asked the Captain.

U.S.S. Tucson Submarine
Sunday 0614 October 21
Approximately 75 feet deep in the Pacific Ocean
150 miles North West Guam

"Time to impact?" asked the Captain. Seaman Trainor knew this question was coming as soon as he hit the fire button the first time and started his computations after the second torpedo left the tube. The Tucson had lost some distance due to the evasive maneuver-the Quick Quiet stop, the crazy Ivan as it was termed. She now was approximately 2500 yards behind Kuái 1. At 46 MPH the torpedo would arrive at the target in 3.5 minutes due to the speed change in Kuái 1. Seaman Trainor completed his calculations and stated, "Three minutes 15 seconds…MARK". It was a long

three minutes. Seaman Marianelli, the SONAR operator, was

busy watching for any defensive moves of the enemy sub and

counting the number of missiles launched. He had counted

eight launched so far.

Kuái 1 Submarine
Sunday 0613 October 21
Approximately 10 meters deep in the Pacific Ocean
150 miles North West Guam

    Captain Wing Ping changed his mind and decided to

fight the enemy sub, which was probably American. "CEASE

FIRE, CEASE FIRE, EMERGENCY DIVE, COME RIGHT HEADING 330

AND GET A TORPEDO SOLUTION READY FOR THE NEW CONTACT. CLOSE

MISSILE DOORS!" Kuái 1 turned toward the enemy and the

captain ordered the fire control seaman to fire when he had

a solution on the enemy. He hit the fire button 6 seconds

later and sent one torpedo out the tube and was almost

configured for another when the first American torpedo hit

its mark. Kuái 1 was hit on the left side about ¾ of the

way down Kuái 1's hull and blew a huge gap in the double

hull at the engineering bay and aft torpedo room causing

secondary explosions. "YEAR OF THE GOLDEN SNAKE!" The

Captain Yelled. The Tucson's second torpedo detonated 3

seconds later in the mass of explosions/implosions that

used to be the Chicoms pride of their navy.

U.S.S. Tucson Submarine
Sunday 0614 October 21
Approximately 75 feet deep in the Pacific Ocean
150 miles North West Guam

"CONN SONAR, WE HAVE AN IMPACT SOUND FOLLOWED BY

BREAKUP SOUNDS. I THINK WE HAVE A HIT! YES, DEFINITELY WITH

OUR SECOND TORPEDO HITTING AND EVEN SECONDARY EXPLOSIONS

AND IMPLOSIONS!" Captain Miles said, "SONAR CONN, that's

great. I hope we didn't just start World War III." Seaman

Marianelli cut the Captain off, "SIR, FISH IN THE WATER,

FISH IN THE WATER FROM THE CONTACT RIGHT BEFORE OURS HIT!"

"COME RIGHT TO NEW HEADING 300 DEGREES, 40 DEGREE DOWN

Bubble AND AHEAD FULL - EMERGENCY DIVE. NAVIGATOR HOW DEEP

IS THE FLOOR?" "335 FEET". "Helmsman at 230 feet arrest our

descent and bring us down another 50 feet". "Aye Aye sir,

arrest at 230 feet and bring us down another 50", said the

Helmsman. "Launch Countermeasures," said the captain.

Captain Miles walked over to the defensive countermeasure

console. He saw the countermeasures launched and asked the

seaman to launch the experimental countermeasures code name Poseidon's Trident from the aft tube as the Tucson was running for her life. The enemy torpedo looked like it had been fooled by the countermeasures but it did not explode. The torpedo moved away from the noise maker trying to find the real target by active SONAR pings. Seaman Marianelli said, "Torpedo back to active SONAR and pinging. She has reacquired us!" "TIME TO IMPACT", asked the Captain. "Two minutes thirty seconds." "Captain new countermeasures ready." "Launch it!" The Trident left the aft tube and looked like a standard torpedo but the differences stopped there.

The Trident was designed by Raytheon to be much like a cluster bomb for the air but under the water. The Trident weapon system was much smarter than a typical cluster bomb. The designers had to account for the weapon to be deployed from the aft tubes like in the current scenario and to not damage the towed SONAR Array. Tests had shown it could distinguish between the array and an enemy weapon but the Captain wasn't taking any chances. "Reel in the Towed Array!" "Aye Aye sir the array is being reeled in".

The Chicom torpedo was tracking the Tucson and was within 350 yards. "Sir, Trident has acquired the torpedo and is initiating its attack" stated Seaman Shand. Seaman Marianelli said, "Sir, I have several small explosions, and no active SONAR pings". "Copy," the Captain said coolly. He thought to himself, *I just spent two million dollars of the tax payer money. Was it worth it?* The whole conn was looking at the Captain as he looked each one of the young sailors in the eye, *Hell yes it was!* "CONN SONAR, enemy torpedo destroyed, sir."

"Helmsman, make your depth 100 feet come right to 135 degrees." The helmsman acknowledged the Captain. "Communications, contact PACAF and tell them they have missiles headed toward Guam and tell them how many were fired," said the Captain. They were finished with this fight until they received new orders.

"XO has the CONN. I'm going to my quarters for a scotch on the rocks."

B2 Spirit of Mississippi Call sign Banshee 2
Sunday 0620 October 21
5 miles East Guam on Final

Capt Ricks was the pilot on this particular mission and he was bored out of his mind. Whoever gave the order to orbit B2s constantly over the ocean was just plain stupid. Either land them at Anderson or fly home. He was on his 6th hour of orbits and was about to land and let his squadron mates, In the Spirit Kansas, call sign Banshee 1, have a six hour tour of the South Pacific. One small joy he had was an electric thermos filled with hotdogs. So far he had eaten three chili dogs with onions, cheese and ketchup. He had long sense mastered the art of eating a chili dog and flying. The aircraft commander was not so talented…yet. He had spilled some chili on his flight suit and engaged the auto pilot. They were both stoked to almost be done with this rotation.

A radio call came in for Banshee 2 to perform a low approach only. Then immediately after the controller issued instructions for Banshee 2, the controller instructed Banshee 1 to taxi into position and cleared for takeoff after Banshee 2 performed the low approach only.

The crew of Banshee 2 looked at each other and said "What the heck?" Then they knew what the heck when they simultaneously saw out the windscreen and heard/saw the Threat/Warn light and horn warning them that missiles had been launched. They saw a ship to the north firing missiles. The Pilot firewalled the throttles and turned on all defensive electronics and pulled up. Banshee 2 was a

bat out of hell. Chili dogs were flying. Only when the

missiles turned to the north did they somewhat understand

what was going on. The order to orbit the base had paid

off.

DDG-178 Japanese Destroyer JS Ashigara
Sunday 0620 October 21
10 miles North Guam

Captain Yamamotto was third generation of sailors in

the Japanese Navy. His father was very proud of him for

rising through the ranks and being the Captain of his own

boat, one of the most advanced warships on the planet. His

father would joke with him after several glasses of sake

that in his day he shot at the U.S. instead of being their

ally.

The ship had concluded some war gaming with the U.S.

the previous week and decided to take some R and R at the

port of Guam. The boat received orders to provide cover for

the island until the U.S. Navy could replace her with one

of their own since they all returned to Pearl Harbor to

replenish supplies and refuel. Capt Yamamotto had heard of

the attack on Diego Garcia and knew there were U.S. war ships now defending that island and looking for possible Chinese subs in the area. An important call had come to the bridge for the Captain from Anderson Air base operations. He took the call and then exclaimed, "Battle Stations-this is not a drill!" Come left heading 270 and prepare to intercept incoming cruise missiles. I want the Phalanx system online and SM2s ready. They were launched 150 miles out about fifteen minutes ago from the north!" They should be inbound very soon!" Exclaimed the Captain. "RADAR has multiple contacts inbound straight for us!" Said the RADAR Operator. "Weapons free, Weapons free, engage at will with all defenses!" Ordered the Captain. His well-trained crew just finished exercising this scenario, the only difference being he had other ships for coverage during the exercise for overlapping fire. That wasn't the case they had now. He knew the classified number of threats he could successfully defend and that number was eight vampires at one time if the ship was the target. He would have thought his ship was the target if he had not received the intel from the Americans about Diego Garcia. The ship was not limited by weapons but by the fire control and the RADAR the ship had.

SM2 missiles were good but not the best. The Americans always saved the best stuff for themselves. The U.S. Navy had the SM3 missiles and they were almost unstoppable. Four SM2s rippled fired from the front vertical launch tubes and were out of sight within seconds. "How many vampires are in the air?" asked the Captain. "Initial count was five and three of those, make that four of those are destroyed" said Seaman Isk. "What about the fifth one?" asked the Captain. "It was on the RADAR and now off but we haven't sent a missile because it is on the East fringe and we haven't locked it up on RADAR yet. Also There are 2 coming in from the west and I'm not sure we can engage those and the ones coming straight in." "CEASE FIRE CEASE FIRE", exclaimed the Captain. He knew the ships missile defense was getting overwhelmed. "TARGET ALL WESTERN VAMPIRES WITH OUR MISSILES AND LEAVE THE OTHER VAMPIRES TO PHALANX!" A quartet of Aye Aye sirs were heard on the bridge followed shortly by a volley of four missiles fired from the deck all heading west to intercept. "Sir, we have two powerful S and X Band RADARs that just came on line from the south, the island I think. "Ignore them, maybe that is someone who can help us."

B2 Spirit of Kansas Call sign Banshee 1
Sunday 0635 October 21

    Banshee 1 was holding short of the active 24R on taxiway K. The Pilot looked out his window to the right and saw Banshee 2 coming in and they knew they were replacing them for hours of fun staring at the Pacific. They also knew that they had probably eaten six chili dogs a piece and there air cycle machine probably couldn't keep up with the natural gas being unleashed in the cockpit. Then they heard the strange abbreviated radio call from the tower and were just as baffled as Banshee 2. When Banshee 2 hit full military power and launched into the sky they figured they should be moving with a sense of urgency as well so they went half throttles into turning onto the active and then full military power for takeoff. Again something caught the Pilot's eyes out his window and at the moment of rotation, when the nose gear lifted off the ground the pilot saw the liftoff of a Patriot missile from a battery north of the base. He exclaimed to the aircraft commander, "Turn south and get away from the island. Vampires inbound, I'm running the combat checklist!" He reached for the checklist and saw an explosion…"Oh man Bustin Nutz storage was just hit. I had my Harley Davidson motorcycle in there!"

Sunday 0635 October 21
Ritidian Eco Beach Resort

Mat Hattier had one fun night! So much fun that his whole body ached as he was swimming in and out of consciousness. He was laying on a wicker chase lounge on the beach that was missing some slats and it had waffled his legs. Apparently he had forgotten to go back to the Cabana to sleep and just wound up here. He opened one eye as the sun light started to come onto his face.

Then he heard the horrible sound of an air horn followed by what sounded like the space shuttle lifting off and he fell off the lounge and got sand all over his face. He looked up in time to see a missile fly 100 feet above his head and headed out to sea. He didn't know what was going on but he did know it was time to leave the beach.

Sunday 0635 October 21
Vacant Lot owned by U.S. Air Force just east of the Bustin Nutz Storage

"Fire Two!" commanded Major Splint. There was only one vampire north east of the island so far, but it was far enough away that the first Patriot Missile might not lock onto it, so they sent a second to back it up. They could always redirect the second missile to another vampire if

the first destroyed its vampire and a second appeared.

DDG-178 Japanese Destroyer JS Ashigara
Sunday 0640 October 21
10 miles North Guam

A wave of three vampires were coming straight in from

the north. One was on a flight path to go directly overhead

of the ship or to hit the ship but fortunately it was mowed

by the mid-ship Phalanx system, which puts down 3000 rounds

a minute of high explosive bullets. Two vampires were

coming in from the west of the ship. Both the fore and aft

Phalanx systems targeted the same missile and brought it

down, but that let one through. That vampire flew past the

ship. The Starboard side Phalanx was not programmed to fire

on a missile that over-flew the ship so it locked on but

the missile was flying away and not toward the boat so

Ashigara missed that vampire. The vampire found its mark on

the island which was targeting the runway where Banshee 1

had been rolling 3 minutes before. The vampire flew in but

missed the runway due to its CEP. It exploded over a parked

B1 Lancer. No personnel were injured but the B-1 caught on

fire and the base commander instructed the fire department

to let her burn until the attack was over.

B2 Spirit of Colorado Call sign Banshee 6
Sunday 0640 October 21

Banshee 6 had been orbiting west of the island by

about 100 miles. The Aircraft Commander made a decision

after the base came under attack to head for Missouri,

Home. He contacted his tanker support and determined they

could get close enough to hit another tanker close to the

U.S. so they bugged out for home.

Sunday 0641 October 21
Vacant Lot owned by U.S. Air Force just east of the Bustin
Nutz Storage

Private Sanders, yelled, "Hit, down one vampire and we

have a clean scope. No more vampires in our sector." "Good

job team! We saved lives today. You set up in record time

and we just got our scopes on when vampires were inbound,"

Said the Major.

45.

B2 Spirit of Kansas Call sign Banshee 1
Sunday 0635 October 21

Banshee 1 also completed the combat checklist and

found Banshee 2 60 miles west of the island near the track

that Banshee 1 would have gone if not for the attack. They

both found the tanker and Banshee 2 maneuvered into

position to take on fuel. She had around half the normal

load of fuel in her tanks whereas the Kansas was full with

just having taken off. As soon as Banshee 2 connected to

the boom from the C/KC-135 they communicated with the

tanker by secure communication through the fuel boom so no enemy could intercept the conversation. The Commander asked what the extent of the attack was and the 135 crew told them two missiles out of eleven struck the island. One hit a B-1 and reduced it to aluminum slag and the other took out a storage facility. The C/KC-135 had satellite communication from PACAF and relayed a mission set to Banshee 2. They instructed Banshee 2 to tell Banshee 1 through hand gestures to connect to the tanker after Banshee 2 and get her mission orders.

Banshee 2 finished tanking and disconnected. They maintained radio silence and decreased altitude and banked right to a heading of 335 for Okinawa Japan. There they would land refuel and get further orders. Ten minutes later Banshee 1 had the same orders.

B2 Banshee 1 and Banshee 2
Sunday 1005 October 21
Kadena Air Base Japan

Banshee 1 and Banshee2 landed at Kadena Air Base and were marshalled into two separate hangers and the hangar doors were shut almost before the engines were shutdown. "We have to go to the commissary and buy a jumbo pack of hotdogs", Said Capt Ricks upon shutdown. The aircraft Commander looked at him like he lost his mind and then said, "Ya let's go."

Figure 12 Spirit of Mississippi in the Hangar at Kadena AB.

46.

B2 Banshee 1 and Banshee 2
Sunday civil twilight October 21
Kadena Air Base Japan

Both Spirits had been refueled and the crew rested as much as they could between mission briefs and preflight. The crew of Spirit of Mississippi had been briefed separately from Spirit of Kansas for Operational Security (Opsec).

As soon as the sun went down the hangar lights were turned off for both hangars and the doors opened. Both Spirits taxied out to the active runway. Kansas would lead and Mississippi would be the number two ship in formation.

Both aircraft would form up out to sea south east of the

island. Kansas taxied into position and went to full

military power, released brakes and started her takeoff

roll. Mississippi taxied into position and held.

Fifteen minutes later they were in formation on a

heading of 225 degrees over the open Pacific.

B2 Banshee 1 and Banshee 2
Sunday 2100 October 21
70 miles South of Taiwan

Both Banshees turned to a new heading of 255 degrees

and flew the new heading for thirty minutes. After that

time Banshee 2 turned to a heading North of 015 degrees.

Banshee 1 turned to a heading of 240 degrees. The Pilot of

Banshee 1 saw Banshee 2 turn off their position lights

which reminded him to run the combat checklist.

Banshee 2 started performing racetracks in the sky at

41,000 feet because she only had twenty-minutes of flight

time to target but had a time on target in one hour at 2230

Okinawa time.

B2 Banshee 2
Sunday 2230 October 21
41,000 feet close to Quanzhou China

    "Put away the chili dogs and run the Bomb drop

checklist", said Major Zillinger, the aircraft commander

for Banshee 2. "Copy", said Capt Ricks. "Arm the Mark 82,

500 pound bombs first and then on the way out we'll use the

deep penetrator on the command and control bunker."

"Yes Sir".

B2 Banshee 1
Sunday 2230 October 21
39,000 feet over Hainan Island

    The pilot, Capt Ware, said, "Bomb drop checklist

complete and computer updated with Inertial Navigation

System (INS) and cross referenced with Global Positioning

System (GPS). We are ready!" "Very Well, DROP DROP DROP." A

few seconds later "40 bombs away, we are clear for egress".

"Turn to heading 045 degrees and let the decoy fly due

south. Time to get back to friendly airspace", said Capt

Valance.

B2 Banshee 2
Sunday 2235 October 21
41,000 feet over Quanzhou China

Banshee 2 flew over the military docks China used to bring in troops and weapons to intimidate Taiwan. Major Zillinger said, "DROP DROP DROP." "Bombs Away", replied Capt Ricks. The bombs released at their pre-programed time initiated by the offensive avionics computer. By the time the bombs were exploding Banshee 2 was already nearing her second objective of the C2 bunker.

In the pre-flight brief the B2 crews were instructed to continue on to secondary objectives instead of performing Battle Damage Assessment on the primary objectives. Another military unit would be doing BDA for the strikes.

Sunday 2230 October 21
Yulin Naval Base Hainan Island "Secret" PRC Sub Base

Capt Wong was getting ready to lay down on his rice mat and go to sleep. His wife, Ling Xu had put his 1 year old son down over an hour ago but he hadn't seen her come in the bedroom after putting him to sleep. *She must have fallen asleep next to the crib,* he thought. Just then the ground shook and pictures fell off the wall and the one-

year old screamed. Ling Xu woke up in a panic and grabbed

the baby. Capt Wong thought a gas main had blown up, which

wasn't that rare on the island, until the second, third,

fourth and on and on explosions hit. He grabbed the family

and got under the kitchen table. He knew they lived 3 miles

from the base and he also knew the American's missiles and

bombs were more accurate than theirs but accidents could

still happen.

The bombs finally stopped. Capt Wong went to get

dressed and his wife was crazy with fear. He had to slap

her to get her to stop clinging to him. He had to go see

the damage. His life's work. *Would the PRC kill me for not*

protecting the base? How did they know where it was?

23 Space Operations Squadron (23 SOPS)
Saturday 1130 October 20 (Sunday 2230 October 21 China
time) Schriever AFB CO

"Why in the hades do we need to be at work on a

Saturday"? Complained Airman Jones. Capt Hanes exclaimed,

"Shut your pie hole Airman and look at your console!" "What

exactly am I looking for in this crappy Chicom fishing

village?" Capt Hanes answered Airman Jones, "I'm fairly

certain that is no fishing village since we received orders

from Space Command to do Battle Damage Assessment for this

area." "OK sir, there is no damage to report…Oh crap, Sir I

have explosions in real time! I've never seen that before!"

The Captain replied, "I wonder who hit them? Taiwan or us?"

"MSgt Johns are you seeing this?" MSgt Johns replied, "Sir

I have a different set of orders for BDA at a different

location. Some little island south of China…I think. I am

looking at Hainan Island. No fireworks yet…wait I am also

getting live explosions and secondary explosions."

Sunday 2230 October 21
Quanzhou China, Chinese Army/Air Command and Control

"Commander Bi we have reports of explosions at the

dock." Commander Bi replied, "Are there any surface

contacts or air contacts on RADAR?" "No sir!" "It is an

attack from Taiwan! The Capitalists are attacking! Initiate

fire control mission Alpha-Full offensive attack. Alert all

missiles batteries weapons free and to defend themselves

against all threats, either seaborne or airborne!"

B2 Banshee 2
Sunday 2230 October 21
41,000 feet close to Quanzhou China

As the 4,700 pound Deep Penetrator came off the right
rack the B2 rolled slightly to the left compensating for
the lateral load being rebalanced from the 40 Mk 82 bombs
that were already released from the left side a few short
minutes ago, Capt Ricks said, "Turning toward friendly
airspace, do you want me to release the decoy?" "I will
release it west with a self-destruct code for 45 minutes.
We'll see if the stick jockeys will buy it.

Sunday 2231 October 21
Quanzhou China, Chinese Army/Air Command and Control

All communication with the C2 bunker went quiet with
several missile batteries asking for guidance due to the
explosions on the dock. No orders to fire on Taiwan were
received. Several days later Commander Bi, or what small
pieces they could find in the smoking crater, were buried
with full military honors.

Sunday 2250 October 21
Quanzhou Jinjiang Airbase

Four Sukhoi Su-30MKK lifted off from the air base at Jinjiang 10 miles west of Quanzhou for combat air patrol and sea scouting. All of the high powered RADARs China had in the area were lit searching for enemies and finding only one. The controller relayed to Capt Ric the general area of the bogey. Capt Ric assigned two of the flight to break right and try to box in the bogey. It was at a high altitude and they would burn most of their fuel to get up to 10,000 meters. Seven minutes later the flight was in weapons envelope of the bogey and they had permission to fire. Capt Ric gave the order and all four fired one missile a piece. Two from the east and two from the south. Thirty seconds later two south missiles and one east missile hit downing the bogey. "THAT'S A WIN FOR THE PRC!" exclaimed Capt Ric.

Sunday 2302 October 21
100 miles North West of Taipei

The Spirit of Mississippi quietly sailed over the Pacific Ocean with no threats and no RADAR signatures on the horizon. AS Major Zillinger checked his MFD for system status he noticed one red light. In aviation a red light is

never good. He pushed the MFD button and the ADM-160B Decoy

System, home screen came up. There was an error message.

The B2 computer lost contact with the drone a few minutes

ago. He pulled up the recorded sensor information for the

last 5 minutes of the drone. There were multiple RADAR

locks on the drone. Including X band search RADAR and S

band airborne RADAR. Then the drone picked up several

missile locks and even evaded one with dispensing chaff.

Then complete data and signal loss. "Rip Little Buddy. You

did your job well." "What are you talking about?" asked

Capt Ricks. "The drone was shot down." "Oooooooh, there

went $80,000. I hope they don't charge us for that. I

already lost a Harley today. Let's eat some chili dogs."

23 Space Operations Squadron (23 SOPS)
Saturday 1330 October 20

    "Colonial Vaughn, We have the classified Battle Damage

Assessment for target designation Foxtrot and Hotel." Said

Capt Hanes over the STU III. Col Vaughn, the 50 Space Wing

Commander, was enjoying a nice day at home with her family

when the classified phone call buzzer went off. Her husband

said, "Well the day just went to crap." She replied, "You

don't know that."

She was totally blindsided by what the Capt was

saying. "Capt…What was your name?" "Sorry Ma'am, I am

Captain Hanes and I am the shift commander for 23 SOPS."

She was taking notes. "Tell me what you have." "Yes Ma'am.

We received flash traffic yesterday morning to perform a

Battle Damage Assessment for two targets. We just did that

today and we went up on two separate supports on Zenit 22

and Zenit 26 satellites. "We caught the attack in real time

and recorded the damage after the bombing." "Captain I am

enjoying a day with my family. Do you have a family?" "Yes

Ma'am, I have a wife, a daughter and a son". "That's

great", said the wing commander, "Why don't you brief me on

Monday or Tuesday morning?" Capt Hanes understood she was

not getting the gravity of the situation. "Because Ma'am

these targets are not some cave in Afghanistan. They are in

China with one being directly across the straights from

Taiwan." Silence, Silence, Silence…OH crap followed by a

string of obscenities. **"I want to see your damage

assessments with your squadron commander in 30 minutes!"**

"Yes Ma'am."

47.

Monday October 8, 2001
Diego Garcia

Col Bolls made his way to the passenger terminal along
with the crusty Major. Lt Craken had his five mobility
bags, personal effects and one SASS sniper drag bag. He
didn't bother going to the armory to turn it in especially
since he was re-deploying to Uzbekistan. Steve was sitting
close to the door when he saw the base commander. He stood
like he was supposed to when a higher ranking officer
entered the room but everyone else didn't seem to notice.

Col Bolls said, "Hey Lt, you aren't the typical dumb lieutenant. You did good getting My birds back in the air to drop bombs. Our availability rate was over 85% and I have been asked to go to the Pentagon for my next tour." Steve replied, "That's great, sir." Col Bolls continued, "I've contacted your SPO commander Col Baxter and let him know what a great job you did here. I also called your next base commander in Uzbekistan and told him you have some cool skills and put you to work. I told him you were especially skilled at supervising the latrine. Well, have a great flight." At that he left and the crusty Major said, "You better hope our paths never cross." Lt Craken looked at him and just said, "Yes sir." Steve wondered if the commander was serious and if he needed to dig out his chem warfare gear to service the Falcon IV remote location latrine when he landed in an estimated 14 hours.

48.

Tuesday October 9, 2001
Aeroport Karshi, Khanabad Uzbekistan

Guppy 1, the C-17 touched down with a jolt that awoke

Steve from his restless sleep. Luckily the C-17 was not

full and Steve was able to rig up his hammock for a better

sleep than the horrible troop seats. Troop seats are made

out of seatbelt material. No one on the flight crew

bothered to tell him they were landing so he was glad

nothing happened or he would have flown through the cabin wrapped in an orange cocoon of nylon fabric.

Aeroport Karshi was a desolate air base. The single runway 25/07 was oriented mostly east/west and was narrow but long coming in around a mile and a half. There was also a dirt strip bulldozed to the south of the main runway but didn't look like it was numbered. C-17s and C-130s can land on unimproved strips but the C/KC-135s and bombers all had to have paved runways.

The local time was 2200 and the C-17 taxied to the ramp on the west side of the base. The Loadmaster came over to Steve and yelled over the noise of the engines, "Lt, when you go down the ramp don't stop moving. They have snipers that like to take pot shots at personnel from the south side of the base. They don't shoot at the airplanes because they know the SPs will hunt them down but there have been a few casualties of members getting off the plane. I think they must be shooting from over 200 yards because they haven't hit anyone that is moving, yet." Steve replied, "Thanks for the warning". He thought about taking his SASS out of the bag and loading it but then the ramp came down and there was nothing but dirt and darkness. The chances he could make out a sniper and engage him before he

got shot were relatively low so he kept it in the case and
got ready to move to cover.

Steve had a small folding roller with handles so he
loaded all he could on it. He had a backpack on his back
and another backpack on his chest and the SASS slung on the
side and took off down the C-17 ramp toward the red roofed
hangar. He overheard the Loadmaster talking to another Lt
that if she were behind the hanger then she would be clear
of the threat. So that is what he did. He estimated it was
50 yards. He was definitely tired when he got there but the
weather was actually a welcome brisk 55 degrees Fahrenheit.
Diego Garcia in contrast was 97 F with 80% humidity.

There was nothing on the backside of the hangar so
Steve and the Public Affairs Lt that he just met were
trying to figure out where to go for billeting or a tent or
a place to get some hot food.

About that time a small three wheeled truck came up
the road they were close to so Steve flagged it down with
his Lackland Laser (a flashlight they issue at basic
training that has a lit cone on the end used to marshal
aircraft (Usually red). It was a local Uzbek national and
he said, "Nima istaysan jack?" So Steve asked him, "Where
is the hotel, Billeting?"…blank stare, "Chow Hall, you know

food?" and he motioned with his hand to mouth. The

contractor said, "Burger King?" Steve was like, "Ya, Sure."

and the contractor pointed north and said, "Mana, o'sha

yo'lga borib, o'ngga buriling. Keyin keyingi yo'lda chapga

buriling. Keyingi yo'lda yana o'ngga, keyin yana chapga

buriling, shunda siz burger kingni ko'rasiz...Burger King"

and at that he left. Steve looked at Lisa the Public

Affairs LT and said, "I guess we go that way" and pointed

North.

They walked a little over a mile and finally saw what

was indeed a Burger King sign in the distance. Steve was

complaining about no base ops van to pick them up or really

anyone. Lisa was just happy to be off the plane and not in

Diego Garcia. She did not like to fly especially sitting in

troop seats.

They made their way into the lobby of Burger King

which was just a metal building that was built recently.

They went over to a table and dropped all their bags off

and went toward the serving line. A Captain said, "Fresh

off the plane I see. Did they try to scare you with the

sniper story?" Lisa replied, "Yes sir, so we ran to the

back of the hangar". The captain Replied, "your more likely

to get run over by the nationals driving those little truck

trycicles they call tiktucks. Lt Craken said, "Right we flagged one down trying to find Billetting. He didn't know any english except 'Burger King'". The Captain said, "I'm Captain Price and I'm from Eielson. I'm the maintenance officer for the "base" (he made air quotes with his fingers)." The two Lts introduced themselves and the Captain was really happy to meet Steve once he found out he could design repairs and do battle damage repair. Captain Price said, "You two get some food and then I'll drive you over to billetting, which is really a tent city but they have heat and AC, so it's not horrible."

Steve and Lisa got in line. Lisa ordred a salad, fries and a diet Coke. Steve ordered two whoppers one large fry and an ice water. They waited for their food to be cooked and then went over to Capt Price's table. Capt Price showed the Lts his base map and showed them on the map where billeting and in processing were located.

The base was not very big so most everyone walked to where they needed to go. The exception was when traveling to the ramp. The ramp was over a mile from the rest of the base and the dirt strip was around two miles if you needed to go look at a damaged aircraft parked out there.

"I can pick you both up at 0630 tomorrow morning to go meet the base commander and then you can walk over to the inprocessing tent. Lt Craken I have a maintenance meeting at 1100 in the big hangar that I would like for you to attend," said Capt Price. Steve replied he would see him at 0630 and then at 1100.

When they finished eating they loaded up in the Captain's Blue bread truck and traveled five minutes to a billetting main tent where they would get their assigned hooches (tents). Both Lts walked into the tent and the extremely bored female SrA looked at them and said, "Ladies first. That is an easy assignment. There is only one tent for females. The good news is there is an enclosed room open and you are the only female officer on base so it is yours." The SrA gave Lisa a map of the base and showed her where the tent was and she was off. "Ok sir, let's see what we have available for you." Said the SrA looking at her laptop. "We don't have any more hard shelters available but we did pitch a brand new tent and we put up sandbags around it. If you hear the claxton sound you run to your shelter. Currently there is only one other occupant in the ten man tent and it has real beds". Steve said, "That will be great. Where is it?" She pointed to the map and steve grabbed all his gear and was off. He turned around and asked, "Who is the other occupant of the tent?" "Let's see. Looks like a Captain Ricks from Kadena Air Base, Med group." She said looking at the registry on the laptop. "Thanks", said Steve as he mumbled, "Med Group" and continued out the door.

It was now approaching midnight local and Steve was tired so he found the new tent that would be home for the next six months and entered as quietly as he could so he didn't wake up the doctor. The lights were on and he didn't see anyone else in the bay so he must be working the night shift. He saw his made up bed at the far end of the bay. Steve picked the bed opposite the door because it was next to a wall. He put sheets on his bed and then his sleeping bag. He halfway organized his stuff and set an alarm for 0530. He then readied his uniform for the next day and opened the case for the SASS and loaded a magazine in the mag well and then hung his 9mm side arm by the holster on the headboard next to his head. He pressed checked the slide to make sure a round was chambered and went to bed.

Steve walk through the front door of Starkville High School and he was looking around. "What time is it?" he said outloud to himself. He then realized he was only wearing bright red and blue Superman underware with those 80's striped socks. The socks had yellow stripes. He thought to himself, Steve, *maybe I have enough time to get to my locker before anyone comes in and at least get some gym shorts and a shirt.*" The school bell sounded and kids started entering the building and Steve started running down the curved science wing and then remembered he graduated over nine years ago and opened his eyes to see his watch alarm going off and blinking...the school bell. "Where in the hec am I?" he asked himself outloud. Captain Ricks replied, "The furthest place from paradise." Steve looked down the tent, blinkked twice and saw Capt Ricks getting ready for bed. Steve said, "Oh hi, I'm Steve Craken, new Battle Damage Repair Engineer." Capt Ricks looked at the SASS he had propped in the corner and said, "That's a nice piece is that what they are issuing these days?" Steve replied, "Some jar head forgot it at Diego Garcia and I talked the armorer into issuing it to me. I guess I forgot to turn it back in." The Capt replied, "Right. Do you know how to use it?" Steve said, "I lit up an OPFOR at the DG attack at 180ish yards." "I heard about the attack. Did you loose anyone you knew?" asked Capt Ricks. Steve nodding his head and said, "Yep our team lost an Electrical troop. She was gunned down next to the base exchange. Two of our team saw it and one guy went full auto on one of the OPFOR and then The MSgt put three in the OPFOR's buddy. Two in the chest and one in the head with his M16. It was a mess!" The Capt replied, "Sorry to hear that." That ended the small talk and Steve got ready for the 0630 pickup.

Lt Craken waited five minutes and Capt Price showed up in his breadtruck. He said, "I'll show you around the base. It's fairly small and you can walk to most places but the dirt strip is geneerally off limits because of the occasional sniper fire. What do you have in that case?" Capt Price pointed to the SASS case that Steve loaded into the truck. "It's a SASS 308 some jar head left in Diego Garcia. They issued it to me because they didn't have any M4s left." Capt Price asked, "Are you any good with it?" Lt Craken replied, "I saved a B-1 from exploding from a chicom's sachel charge by taking him out at 170 yards at the battle for DG." Steve didn't mention it took two shots. "If I know the commander, and I think I do, you won't have to clean the latrine. You'll be attached to the Security Forces detail for counter sniper ops for your additional duty. We don't have anyone here that knows how to engage a target past 100 yards and that is even iffy-really only about 50 yards."

The base tour did not take long as they pulled past some concrete barriers called Texas Barriers because they were the largest barrier. They turned into the make-shift headquarters that used to be the headquarters for the Uzbek Air Force. Steve thought that was weird seeing old Mig 25s in various states of disrepair on the flight line.

Capt Price went into the mainteneace meeting first and sat down at the table. Steve remebered what Jesus said about taking the most important seat at the banquet...even though this was a meeting he sat along the wall in the non-important seats. Truth be told he really didn't want to sit at the table. The exec called the room to attention when the commander, Col Smith, came in. He said good morning to everyone and Capt Price spoke up and said, "Sir we have a new officer, Lt Craken, assigned to us from, recently Diego Garcia and his home station is..." Capt Price looked at Steve and Steve said, "Oklahoma City, Tinker, sir." Col Smith said loudly, "All there is in Oklahom are steers and..." About that time the Executive Officer cleared his throat because the commander was about to say something that wasn't Politically Correct and he already apparently had offended someone on base becasue he had an Equal

Employment Opportunity Counsel (EEOC) complaint. Steve

chuckled and replied, "Steers are from Texas and I'm not

from either." About that time the commander started the

meeting but Capt Price interjected, "Sir, Lt Craken is an

Aircraft Battle Damage Repair Engineer and he saw some

action in the Battle for DG against the Chicoms." At that

everyone in the room stopped what they were doing and

looked at Steve. The commander said, "Is that right?" and

looked at another officer at the table. "Jim I am assigning

Lt Craken to you to augment your Security Forces." Maj Jim

Forstal, replied, "Yes sir, I'll get him spun up right

away."

Lt Craken said, "That will be fine, sir. Do you know

when my ABDR Team will be here from Tinker"? Steve got wind

that HQ was spinning up another team to send to Uzbekistan

because the C/KC-135 team was staying at DG and going home

in a few weeks back to Tinker. The commander said, "I had

no idea you were coming until yesterday when Col Bolls

called me and told me you were coming and you would be a

great addition to the latrine crew. I thought you were

civil engineering. We have nationals that love cleaning the

latrines because we pay them a year's worth of wages in two

months. I guess you really pissed him off". Steve replied,
"Sir, we didn't see eye to eye on how to repair aircraft
and realistic schedules." "Don't sweat it Lt, I know Col
Bolls. We were at Air Command and Staff College. He is a
jackass. As long as you fix jets and keep that sniper at
bay we'll get along fine. Having said that I do not know
when an ABDR team will get here." Steve said he had a
presentation if the commander wanted it on what ABDR brings
to the fight. The commander said he was fairly busy but
maybe in the coming week he could find time. In the mean
time Capt Price would work with him and the backshops to
fix aircraft.

The meeting lasted an hour with typical mundane info
and various flight line and back shop commanders briefing
the base commander on status and sortie generation. It was
good for Steve to learn who the commanders were so he could
talk to them after the meeting and get their buy-in so when
he met with the shop NCOs they would help him get materials
or provide troops to help repair damaged aircraft. At the
moment Lt Craken was a one man ABDR team!

Tuesday October 9, 2001
Aeroport Karshi, Khanabad Uzbekistan

Lt Craken was loaded down. Not as bad as the redeployment but he had his ruck sack on, engineer's kit and
SASS drag bag along with his normal load out of 9mm and
extra mags. He was walking around the base looking at
damaged aircraft.

During the maintenance meeting with Capt Price this
morning he discovered there were two jets with various
damage; One B-52 that had significant battle damage caused
by Anti-Aircraft (AA) fire close to the base during the
landing phase of their mission. The whole Security Forces
team was dispatched when the B-52 crew reported the AA
close to the base. The base SF team annihilated the OPFOR
running the mobile AA. That was two weeks ago. They parked
tail number 59-2596 way out in the dirt close to the dirt
strip.

Steve wanted to get out to the B-52 first because he
heard there was bad weather coming in later in the day and
he'd rather not be out in a blizzard hiking around. He also
wanted to see if he could see any obvious sniper hides in
the topography around the base.

His primary mission was ABDR so he went to inspect the B-52 first. He opened the crew entry door, turned on his flash lite and climbed in. Since the B-52 was parked in the dirt there was no Aerospace Ground Equipment (AGE) to provide power/hydraulics/air, nor did Steve really know how to use it. Back when he was a jet Mech he had a tutorial on the AGE equipment and basically the AGE Tech said, "Don't touch it if you are not an AGE tech."

He was able to find a crew member on the flight when it was hit before he headed out to the bird. The crewman told Steve there was a loud noise and shudder from the left side fuselage. From Steve's external inspection he did find skin damage around the bomb bay and wanted to gain access to the inside of the bomb bay to see what internal damages there were. The skin repair didn't look like it would be bad to design or install by Steve. A B-52 has a unique internal structure. No other aircraft in the U.S. inventory is like it. For comparison the C/KC-135 has several hat stringers running the length of the fuselage all the way around the internal circumference. The B-52 does not have the small stringers running the length of the fuselage. Instead has four massive longerons with a cross section of a box (rectangle) with a 12 inch width. The skin of the B-

52 actually takes all the shear load generated from all the

loads applied to the body skin (bending, torsion, etc.).

That is why when you look at the fuselage you can see waves

in the skin (See figure below).

Figure 13 B-52 nose and fuselage with wavy shear skin.

Steve crawled to the rear of the lower cockpit and

opened the pressure door that led to the bomb bay. There

was a small catwalk above where the bombs would be located

if the B-52 were loaded. Thankfully, she had dropped her

entire load before returning to base so no bombs were in

the bay when the aircraft was hit.

Steve's fears were realized when he saw the bottom left longeron with several small holes through both sides and one gaping torn hole in the bottom of the longeron. Several hydraulic lines were cut clean through and must not have been an issue with landing because the crew did not report any unusual landing issues like lack of brakes or flight controls. *Apparently, the redundant systems worked as designed*, thought Steve. Steve started thinking of how to repair the damaged Longeron. He was thinking he may have to do a very ugly and very large scab patch over the longeron to take up the load that was missing from the original structure. Maybe he could find some steel sheet or angles instead of Aluminum.

Steve was thinking *we'll get her airworthy and fly her home to depot for a permanent repair*. That would be Similar to the exercise they had back at Tinker. *I'll write up no bomb load, minimal fuel, minimal crew, no known turbulence for the flight restrictions*. The B-52 will have to wait on the other bird's damage. He made some rough sketches and took measurements of the damages so he could make some calculations.

Steve exited the B-52 and buttoned up the doors. He thought to himself, *now to survey the surroundings for possible sniper hides*. He put his ruck behind the landing gear of the B-52 and dug out the spotting scope, rangefinder and a notebook. Steve used the notebook to sketch a range card of the south side of the base which was mostly high plains desert. There were two possible locations that would make a good hide. Steve ranged both positions with the range finder. The first location was a rock knoll. It ranged out at 300 yards from the B-52. The second position was some old metal buildings that looked like they used to be for the fire department but in this hemisphere who knew what they were built for. The two looked old. He made a note to see if there was a way to get on the roofs of the metal buildings for later when he checked them out with the SF Team.

SSSHHHHHHffffff…ping…BOOM. A round zipped past Steve's head and ricochet off the landing gear stainless steel strut. Steve ran and slid behind the B-52 landing gear for cover and concealment. *That round hit a half a second before I heard the boom…That came from the buildings! Steve came to the realization*. Steve grabbed the rangefinder and quickly ranged the buildings. 502 yards. He ripped open the drag bag and grabbed the SASS, slammed a ten round magazine in the mag well, charged the rifle with the "T" handle, turned the selector to "Fire" and extended the bipod legs.

They are scanning for me and adjusting their optics. There is probably no way they can resolve me (not enough magnification to discern between the B-52 landing gear and a human target) lying down behind the gear…that's a big PROBABLY, He thought. *I should call this in…after I kill them, but if I miss we could set a trap when/if they run.*

"Security Forces, Engineer One", said Steve as he keyed the radio mic. "Go for Security Forces-SF Tomcat is my call sign", Came the reply from the A1C manning the radio at the SF Ops tent. "I am taking sniper fire on the dirt ramp next to the parked B-52. Are you ready to copy a "SALUTE" report?" "Go for SALUTE Report" came the reply from SF Tomcat.

SALUTE Report

Size of Unit – Unknown but probably one shooter and one spotter.

Activity – Currently engaging me beyond 500 yards.

Location – Unconfirmed but probably abandoned metal buildings southwest of the dirt strip.

Uniform – Unknown, I'll get the spotting scope on them in a few.

Time – Now, 1317 Local.

Equipment – Sniper rifle.

"I am requesting SF support on the road to the West of base to capture sniper team. I am going to engage them and I think they will run since they have not been challenged prior, over," replied Steve. The A1C immediately came back, "Standby one, Engineer One". "Copy SF Tomcat."

While Steve waited for the go on the trap he looked up the ballistics table taped to the side of the rifle butt stock. For 500 yards the table read 23.2 clicks up (2.3 MRADs up). You can't do a 0.2 click so that just gave the shooter information that the aim point inside the scope needs to be slightly higher than the center of the cross hairs. There is a 41.9 inch drop of the NATO bullet at 500 yards for the ammunition Steve was issued. Steve clicked

the elevation turret up on the SASS scope 23 clicks and

adjusted the magnification from 5 times to 15.

SSSHHHHHHffffff…BOOM. The sniper round hit the dirt

about 20 yards from Steve's hide behind the B-52 landing

gear. *Not very accurate. They are probably shooting a very*

old Dragunov or maybe even a Mossin Negant, thought Steve.

Both of the Russian rifles that shoot the 7.62X54 mmR

Cartridge. The 7.62X54 mmR Cartridge is slightly bigger

than the NATO cartridge but the rifle they are using

probably has been used so much that the barrel riflings are

shot out making the accuracy poor, explaining why they

haven't scored many U.S. kills. It was mostly harassing

fire.

Steve slowly crept a little forward of the B-52 right

front tire on the right front landing gear truck. He peeked

around with the spotting scope set on 20 times power. He

acquired the buildings and looked at the roofs and saw

movement. He put the scope away and pulled the SASS over to

his position. He was prone with most of his body behind the

gear truck. His legs were spread and feet sideways against

the ground giving him the optimal shooting body position.

The radio crackled to life, "Engineer one, SF Tiger", Steve thought, *SF- Tiger…That must be the SF commander.* Steve answered the call while looking through the SASS Scope, "Engineer One go." SF Tiger replied, "Let me get this right, you are currently under fire from the enemy, beneath a multi-million dollar bomber and you want us to roll HUMVEEs to the road on the West side of base to cut off retreating OPFOR?" Steve replied, "Affirmative, sir." I'm Ok with you being bait. We are rolling!" Hopefully the OPFOR wasn't listening to the radio because they didn't have secure comms.

"SF Tomcat and Net (any other friendly on the net listening that needed the info), Engineer One Update SALUTE Report." Steve said. SF Tomcat came back with, "Go for update SALUTE."

Updated SALUTE Report

Size of Unit – 2 FAMs (Fighting Age Males)
One shooter and one spotter.

Activity – Currently engaging me beyond
500 yards.

Location – Abandoned metal buildings
southwest of the dirt strip.

Uniform – Black pajamas.

Time – Now, 1321 Local.

Equipment – one Dragunov and one AK-47.

"Copy updated SALUTE Rep…," replied the A1C. "Engineer
One, SF Tiger," The SF Commander stepped on SF Tomcat's
radio call. Steve answered, "Go for Engineer One." "We are
in position and do not have eyes on any OPFOR." "Copy I
have eyes on you and OPFOR. OPFOR does not have a clear
line of sight to HUMVEES. Standby. Getting ready to send a
round. I am engaging," replied the Lt.

The Lt took in a full breath, let out about half of it
and held the rest. The SASS was already positioned to make
the shot and was solid on the bipod and Steve's shoulder.
His right hand was on the pistol grip and his trigger
finger moved from the side of the lower to the trigger. His
left arm went under the rifle and his left hand rested on

his right bicep. He lined up the cross hairs dead on the spotter because there was an "I" beam in the way keeping him from making a kill shot on the shooter, which let Steve know the shooter could not hit him from his current hide.

Lt Craken slowly squeezed the trigger and the rifle fired with the recoil going into Steve's shoulder. The 168 grain bullet left the muzzle at 2650 feet per second and a half a second later hit the OPFOR spotter just below his neck killing him instantly.

Steve got back on target to see if he could send another shot at the sniper but their hide was now empty. "One OPFOR down and one squirter," Lt Craken said on the radio. "Copy we surrounded the OPFOR and we now have him in custody. We'll send a HUMVEE to pick you up." replied SF Tiger. "Copy, I'll be ready Engineer One out," said Lt Craken.

Steve stood up and shook the dust off. That was the second time he had been shot at in his life. The first time…

Thursday 0230 July 4, 1991
B Quick Gas station Old Mayhew Rd. Starkville MS

It was early morning and the two teens had been
harassing a firework stand owner. He went by the name
"Lizard". So Steve, Best-Friend Kyle and some other friends
went to buy some fireworks for the Fourth. This deranged
nut job, Lizard, starts yelling at them to leave. They left
but that was three days ago and they decided to go back and
taunt the dude. This was one of Steve's not so bright
moments. Best-friend Kyle and Steve decided to sneak out
the morning of the fourth. They were supposed to pick up
two other friends but for whatever reason they could not
go. So Kyle was driving his red Dodge Dart that had the
white stripe on it just like the old TV show Starsky and
Hutch car. He drove down highway 82 to the gas station and
pulled into the parking lot. The firework stand workers
came out and Kyle drove onto Old Mayhew Rd and stopped.
Steve had his window rolled down and was taunting Lizard.
Lizard drew his 357 Magnum revolver and fired it into the
air. The two teens had watched too many movies and didn't
know what a real pistol gunshot looked like so they thought
it was just a firework. "I'll kill both of you!" yelled
Lizard. "Go ahead," yelled Steve back. Lizard took aim at

the Dart and fired. **BOOM and glass breaking.** Steve had

glass blown into his hair and the back of his head. "Go,

Go, Go!" yelled Steve to Kyle. He punched the gas and lit

up that old straight six under the hood. They drove to the

police station but the gas station was outside city limits

so they had to go to the Oktibbeha County Sheriff's Office

north of down town. The Sheriff did not believe them. He

said, "Maybe he shot a slingshot at the window." Then the

Deputy told the teens to load up in his cruiser and they

were going to get to the bottom of this. They almost hit a

parked car because the Deputy was looking down at some

paperwork in his seat. He yanked the wheel just in time to

miss it. Then they passed a semi-truck on Highway 82 on the

left side and almost ran a car off the road.

The Deputy made it to the gas station without further

incident. "Wait in the car. I'm going to see what he has to

say. The deputy walked up to the fireworks tent and here

came Lizard walking out. "These boys say you shot at them."

Said the deputy. "DANG right I did!" while pulling out his

357 magnum. The deputy put his hand on his service weapon

but did not draw. He exclaimed, **"Put that gun back in its**

holster right now!" Lizard complied. "You need to come to

the sheriff's station in fifteen minutes," Instructed the

Deputy. The Deputy drove the teens back to the station and they pulled up to see the Sheriff taking apart the Dart's dashboard. The bullet had broken the backseat window behind Steve's head, through the car, in between Kyle's head and Steve's head, hit the car's front left window post and broke in half. Half of the bullet went out the driver's window that was rolled down and half went into the dashboard that the Sheriff just found.

The Oktibbeha County Court convened two months later for four hours and the teens never did know what legal charge Lizard faced but they learned that he had to pay a fine for attempting to kill them. The fine was $50 for attempted murder. Kyle joked that his life was worth $25 and so was Steve's. Steve knew better. He knew his life was worth $49 and Kyle's $1.

Tuesday October 9, 2001
Aeroport Karshi, Khanabad Uzbekistan South of dirt strip

One HUMVEE picked up Steve next to the B-52 and the other drove up to the metal buildings. Steve got on the radio, "Net, Engineer One, Does this base have any EOD personell assigned?" Engineer One, SF-Tiger, "No EOD on base." "Engineer One, Net, stay out of buildings due to possible IEDs." Lt Craken looked at his HUMVEE driver SSgt

Wilkes and asked him to drive him over to the rocky knoll

east of their position. They drove over and the Lt asked

him to stop about 50 yards from it and got out to look for

signs that the sniper team had been there. Steve took the

SASS off his should and put the scope up to his eye to look

the area over for possible IEDs or tripwires. He didn't see

any so he moved up and circled the rocks. From 25 yards out

it did look like a great hide with clear lines to the

parked B-52 that Steve was pinned down at and the rest of

the ramp and runway. The ramp and hangars were quite a

distance away, like over 800 yards so they probably weren't

targetting those areas. The parking area and cargo off

loading area was only 300 plus yards away and definitely a

target. There looked like some sketchy stuff around the big

boulders so the Lt did not approach and went back to the

HUMVEE.

The whole team went back to the ops center and the

OPFOR sniper earned himself a black bag over his head and a

trip to the holding tank.

"good job Lt." said Col Smith. "I heard the whole

exchange. You are one cool operator." "Thanks, Sir, I'm

going to check out the other damaged bird and see about

designing repairs for it if I'm not needed around here."

"Go ahead, Lt. I'm sure we have our hands full with the

OPFOR you helped capture." "Oh one more thing, Sir, we

should hit those buildings with a bomb and the rocky knoll.

I think that would end the sniping because there are

definitely no other hides within range of the base if we

level those two areas," instructed Steve. "We can discuss

at the next week's staff meeting," replied the commander.

"yes sir."

Lt Craken left the command Building and walked accross

the street to the hangar. The hangar contained one C/KC-135

with the right wing leading edge fairing removed. A few

maintenance techs looking up scratching their heads on how

to fix the bleed air duct that was mangled on the ground

next to the landing gear. Lt Craken asked the senior

maintainer, "What happened here? Bird Strike or enemy

fire?" "That would be a 20 pound vulture strike." Replied

the MSGt without looking at the Lt. Steve replied, "The

repair for the leading edge is in the -3 Tech order. Have

you ordered a new bleed air duct from supply?" The MSGt

stopped gauking at the wing because usually no one around

here knew the tech orders or how to get parts. "Oh sorry

sir I didn't notice you walked in." The Lt recognized the

MSgt as the supervisor of the structures shop, MSgt Powell. "No worries," replied Steve. MSgt Powell said, "Supply laughed at us when we asked for the part. They said that duct hasn't been procured in thirty years." "I guess we'll have to fix it." Exclaimed Steve. "I heard what you did out in the dirt. Heck the whole base has but fixing that duct will be a miracle," said the MSgt Powell. "You supply the muscle, I'll supply the brains,." Steve said smugly.

The starter/bleed air duct is a 6 inch diameter stainless steel pipe that connects the air accumulator to the right side engines to start the engines on the ground. Also the pipe has a secondary function. The pipe supplies engine bleed air to the air cycle machine for climate control and pressurization of the cabin. Most importantly it was used to spin the engines up to 60% RPM to start them.

The Msgt walked over to the pipe and called over A1C Walker. A1C Walker was a huge body builder from Nebraska. "A1C Walker this is Lt Craken. Do what he says to fix the bleed air duct." "AAH, Ok MSgt." Said A1C Walker.

Steve started explaining how they were going to fix the bleed air. "First we are going to make a special tool. Have you ever seen a slide hammer that they use to repair auto bodys?" "Oh ya, I have. I can make one at the structures

shop." Said A1C Walker. "Great make sure it is at least

five pounds with ten pounds being better because that is

stainless steel. Then we'll chain the duct to the hangar

pole and weld the slide hammer to the pipe and bang it out

and then cut the hammer off and move to the next area."

Described Steve. "Ok I got this sir,." said the A1C. "Great

I'll check in tomorrow and see if there are any snags."

Said the Lt.

"Engineer One, Base Ops." "Go for Engineer One." Steve

keyed the radio mic. "We have an inbound F-15 with small

arms damage to the wing, Please meet the aircraft on the

ramp for Battle Damage Assesment." Replied Base Ops. Steve

keyed the mic again, "On my way."

Tuesday October 8, 2001
145 miles south east Aeroport Karshi

Talon Two with Lt Wilson as Pilot in Charge was

crusing at 10,000 feet on the way back from supporting ODA

595 to the south. There mission was to drop bombs from

their F-15E at high altitude. They did so but when they had

released all their ordinance the ODA team requested they

drop down and do some battle dmaage assesment. They responded but came under fire from a twenty two man unit of the Talliban that were waiting for an aircraft to do a low level run. When Talon Two flew below the canyon the Talliban shot at the aircraft with Rocket Propelled Grenades (RPG). At least twenty RPGs took flight towards the F-15E. RPGs are not designed to shoot down aircraft but when waging gorilla warfare you use what you've got. Since the RPG is a dumb weapon with no seeker heads the F-15E sensors did not pick up the launch, except for the IR signature of the rockets, since there were so many of the small rocket motors. The canyon was narrow and one RPG actually hit the left wing at the aileron root. The F-15E flight computer accounted for the damage. There was barely any percieved loss of control of the F-15E due to the RPG damage.

Talon Two pulled up high, rolled level away from the canyon, then turned back and started engaging the taliban with its Vulcan cannon sending some of them to Allah. Talon Two knew they missed some but they radioed to the ODA 595 team that they might have damage and were egressing the area. Then they gave them the battle damage report they initially requested. "Battle Damage Assesment on enemy; two

mobile rocket launchers completely destroyed, one Armored
Personell Carrier (APC) destroyed and several ground troops
hit. Over and out." "Thank you Talon Two. God speed home."
Said the ODA 595 redio operator.

Tuesday October 8, 2001
145 miles south east Aeroport Karshi

Lt Craken heard the F-15E before he saw it. It sounded
normal so that was good. Both engines were probably
functioning normal and only one fire truck was called out
to meet the stricken jet. He didn't think the damage was
severe. That was a good thing because the F-15E is a mostly
composite aircraft of which the Lt's expirence was limited.
He knew of aluminum honeycomb flight controls that the
aluminum skin was bonded over the aluminum honey comb core.
He had repairs for that kind of damage but no repairs for
carbon fiber. If that were the case it would have to be
ferried flighted home or technicians brought in to repair,
or blown up out in the dirt if the other options weren't
feasable.

F-15E landed normally, extended its air brake and
pulled onto the parking ramp and shut down. The Pilot and

the Navigator climbed down from the cockpit and Lt Craken

met them by the left wing. Steve introduced himself and the

flight crew were glad to see him. "Can you fix this?" was

their first question. Lt Craken replied, "I don't know yet.

The only composite repairs we can do here is to replace the

damaged composite with metal. Can you tell me what

happened?" asked Steve. "Sure, we just dropped our load and

the ground team requested a BDA so we rolled into this

narrow canyon and I got this weird alert on the TWR. Then

we heard an explosion and I looked back to see smoke and

dust coming off the aileron," Said the Pilot. Steve asked,

"TWR is your Threat/Warnning Receiver?" Yep was the reply.

Steve followed up, "Any control issues after the impact,

fuel leaking or hydraulic issues?" The Pilot answered, "Not

that we could tell but this F-15E has a computer upgrade

that can sense anomalies in flight controls and counter

with other coupled flight controls. It's really cool stuff.

So when it first hit, rolling was wicked sluggish but

apparently a couple of iterations later the computer

figured out how to compensate for the lost aileron. It was

a smooth ride all the way back here. Wasn't even really

scary," Replied the Pilot.

Steve got on the radio and requested the F-15E towed into one of the hangars so he could start assessing the damage. Ops said they would have the bird in Hangar 3 within an hour because a blizzard was coming.

Tuesday October 9, 2001
Hangar 3 Aeroport Karshi, Khanabad Uzbekistan

Lt Wilson asked Steve, "Where can we get some grub?" "Ahh burger kingni...Burger King yep it's that way about a half mile. Look for the metal building with the blue roof. Billetting is a block over. It has a sign in front of the tent." "Thanks we'll drop by later," said Lt Wilson. "No need the'll probably put you up in the same tent as me. Later," said Steve.

Steve rolled a maintenace stand over to the F-15E left wing trailing edge and climbed up. There was a hole clean through the aileron and the actuator control rod was mangled. The rod was so magled that it was alomst broken in two. Whoa! *That flight crew was lucky. If that rod was completely broken then the aileron would have severe flutter and would most likely departed the vehicle with possible loss of control to the aircraft. Looks like I can design a repair based on the C/KC-135 ABDR Tech Order for*

its aileron. They are both aluminum honey comb with face

sheets bonded to them. I can make aluminum skin sheets with

an internal stiffner attached and then sandwhich the

aileron between the two face sheets using jo-bolts.

Steve found a back shop with an empty desk where he

could write up the repair. After he did so he walked over

to the structure shop to see if he could beg, borrow or

steel some man power to fix this jet. He also thought about

calling the F-15E program office at Robins AFB Ga. He would

need to touch base with them. He put that on his "To Do"

list.

Tuesday October 9, 2001
Structures back shop Aeroport Karshi, Khanabad Uzbekistan

The structures back shop was located off Hangar Two

and it was massive with all the machines one would expect

at a regular base structures shop. Lt Craken walked in and

no one noticed so he went shopping for repair materials. He

found several racks of aluminum sheet and extrusions

including the "T" extrusions he just called out on the F-

15E repair. *Now to look for some steel sheet, angles or*

extrusions, Thought Steve. About that time a SrA noticed
some random person going through their stock and challenged
the Lt by yelling the challenge word, "Mt Olympus!" Lt
Craken yelled back, "Stands Tall!" So the SrA at least knew
Steve was supposed to be on the base. The shop didn't get
any visitors so he walked up to Steve and noticed he was an
Officer. "Sir, can I help you?" Steve replied still looking
through the metal racks, "I'm a Battle Damage Repair
engineer and I need some materials and manpower to repair
the F-15E that landed today and the B-52 out in the dirt
lot." Then he turned around and directly asked the SrA if
he could cut metal and buck blind rivets. "I'm the best in
the shop replied the SrA!" Steve seriously doubted that but
whatever. MSgt Powell walked up and Steve asked if he could
borrow the SRA Tidwell for a few days. MSgt Powell looked
like a toddler dad when the babysitter showed up.
"Absolutely, wait, who will we get to sweep the floors? Oh
never mind. I guess it might keep him out of trouble." Lt
Craken gave the SrA his shopping list for the F-15E and had
him meet him at Hangar 3 with his tools. Steve laid out the
repair for SrA Tidwell which seemed to be an OK kid...kid
he was only five years younger than Steve. SrA Tidwell
started fabricating and Steve said he would check back with

him in a few hours and to NOT cut or drill on the aircraft until he got back except for the damage removal of the skin and honeycomb.

The time was getting late in the day and the storm had indeed rolled in. Steve went from Hangar three to Hangar two to check on the progress of the bleed air duct for the 135.

Steve walked into the quiet Hangar and assumed everyone was at chow until he heard a bunch of profanity from the side of the hangar. He looked over to see A1C Walker with two seperate pieces of the duct work. He walked over and A1C Walker appologized for his profanity. Steve said, "Don't sweat it. So we can weld those back together. How is the rest going?" A1C Walker replied, "Most of it is done so if I can weld these back together then it should be OK. I can't get the pipe completely smooth. It will have some wrinkles in it like a crushed pop can. Will that work?" Steve replied, "I'm a structural engineer not mechanical systems but we'll slap it on and see if we can start the engines whenever the leading edge is repaired." A1C Walker said, "Sounds good."

Wednesday October 10, 2001
Hangar 3 shop Aeroport Karshi, Khanabad Uzbekistan

   Lt Craken started makiing his rounds and started with

the F-15E in Hangar 3.

Figure 14 Talon Two in Hangar Three for repairs.

The SrA had cut out all repair pieces and only needed some

minor rework to break all the sharp edges. Steve laid out

how to install the repair with fay surface sealant and how

to drill the holes for the rivets. So he felt fairly

confident the SrA could install the repair without issue.

He told him if he had any questions to get him on the radio

and he would come over.

   Steve also inspected the push pull rod the SrA made.

Just like the one he made in the lab.

Figure 15 Push Pull Rod for F-15 Aileron.

He left the hangar and went to HQ for the maintenance meeting with the commander. Steve walked in and sat in his usual spot on the wall. The commander walked in and the room was called to attention. Then Col Smith said, "Lt come to the table. We have lots to discuss." "Yes, Sir" Steve replied and moved to the table. Steve noticed Lt Wilson was sitting next to the opposite wall. The commander asked, "Do you have an update on aircraft Lt?" Lt Craken spoke up, "Sir the aileron patch on the F-15 is being installed as we speak and should be ready for a test flight tomorrow afternoon. Lt Wilson said a big, "THANK YOU!" Steve looked

at him and gave him a Lt to Lt head nod. Then Steve said to

the group, "Without maintenance, Pilots are hitchikers with

a cool jacket." Everyone had a good laugh at Lt Wilson's

expense. Lt Wilson said, "True that but thanks still. If

you ever need anything let me know." Steve replied, "Well

there is one thing. It would be nice, if during your check

flight you would be so kind as to drop some Mk 84s on the

two buildings and rock formation south and west of the

flight line." Lt Wilson said, "Um OK sure." Steve

explained, "That is where the snipers like to hide." Col

Smith interjected, "That may be a possibility. Let me check

with legal and the SOFA - the Status of Forces Agreement to

see if we can destroy those." Steve resumed his brief, "The

C/KC-135 leading edge repair is complete and we should have

the bleed air duct repaied as well. We will have to test

the duct to see if there is enough airflow through the duct

to start the engines. I think it will work." The commander

said, "Can't you do some math and tell if it works?" Steve

replied, "I'm not an fluid dynamacist. I could put some

numbers in an equation but it would be an educated guess.

We would still need to test it. I'm fairly certain it will

work. Moving on to the B-52 at the dirt strip. I can repair

that with some steel from the structures shop or actually
borrowed indefinitley from Civil Engineering. When the F-
15E is moved out of the hangar I would like to move the B-
52 in and start repairing it." "Sounds good. Good Job Lt."

Wednesday after dark October 9, 2001
Burger King Aeroport Karshi, Khanabad Uzbekistan

The claxton sounded as Steve was eating a burger. He
grabbed his gear and started jogging to the shelter like
everyone else. He heard on the radio there was sniper fire
coming from the south. He changed directions and went to
the tallest hangar, number three. He went inside and
climbed the stairs to the second level and then found the
ladder to the roof. He climbed up to the roof, opend the
access dor and climbed out onto the roof. He started to low
crawl accross the roof with the SASS drag bag using it like
it was intended...by dragging it. He found an
airconditioner and he kept the large metal box between him
and the sightline to the metal buildings 700 yards away.
The hangar was about 200 yards from the parked B-52 so
Steve knew the range to the buildings and rocky knoll but
he didn't know which hide the OPFOR sniper team was using.

He waited for another shot to see the flash from the barrel. He knew they weren't shooting suppressed because of the distance and he heard the previous shot.

"Engineer One, Saber Tooth." *What is up with these cat call signs,* thought Steve. "Go for Engineer One." "We have sniper fire from the south are you in a position to return fire?" "Affirmative."...**Boom,** Another shot from the OPFOR sniper.

"Saber Tooth, Engineer One, Stand by one. Steve dialed his scope up to 44 clicks. He was issued a new night vision monocular that he attached to the SASS scope already in place. He really couldn't tell what the wind was doing so he left the windage alone. He lined his shot up, took a full breath in and let out half. He had two targets like before in the buildings. He squeezed the trigger and the rifle fired. He had enough time to get back on scope to see the round impact right by 8 inches. He quickly dialed in three clicks left on the windage for 3/10 MRAD or 7.5 inches and lined up his shot. Nothing there. They left. Steve held on target and a couple of minutes later he saw a head poke up and then back down. *Ahh I just need to wait,* thought Steve. Whack *a mole.* Another head popped up and stayed a bit longer. Steve was willing the OPFOR to think,

Thats's it. The American shooter is gone go on back to shooting. Then Steve got back on the radio, "Saber Tooth, Engineer One, there is still a threat but they look fairly comfortable so stand by."

Two heads came up and Steve Identified the sniper. He went through all his firing prep mechanics. He squezzed the trigger...**BANG** from the SASS. He got back on scope ASAP in time to see the sniper recieve a head shot. He aimed for the spotter and fired again but when he came back on scope he saw him running. *Crap miss!* "Net, Engineer One, all clear, all clear. SF come pick me up at hangar three to secure the area," Steve called out on the radio.

Thursday October 10, 2001
5,000 feet above Aeroport Karshi, Khanabad UzbekistanThe F-15E was repaired and lifted off the runway in full afterburner.

The jet climbed up to 5,000 feet to check full maneuverability with the repair. The F-15E was loaded with two AIM 120 AMRAM air to air missiles, two AIM 9 sidewinder missiles, two fuel bags and four MK82 500 pound bombs. The F-15E had full range of maneuverability and no obvious issue from the repair scab patch. Lt Craken would have to inspect the patch once the F-15 landed. The pilot radioed

the tower and asked for clearance to engage the two sniper

hides with two each 500 pound bombs. The tower came back,

"Cleared Hot."

The F-15E rolled into attack on the buildings first.

He dropped two bombs and hit his mark. The buildings were

reduced to slag. The tower came back, "Direct hit on the

buildings." Lt Wilson rolled in to the rocky knoll and

dropped his bombs. One went wide and took out an acient

cattle trough but the other hit its mark and threw smaller

rocks into the air. The tower again radioed the F-15E,

"Good hit. stand by. The runway has to be cleared of

rocks." Fifteen minutes later the runway was clear and the

F-15E landed. Lt Craken met the jet with SrA Tillman to

inspect the repair. It was solid. The Lt looked at the SrA

Figure 16 Talon Two, F-15E taking off for check flight.

and said, "Great job. You just made your first battle

damage repair."

Friday October 11, 2001
Aeroport Karshi, Khanabad Uzbekistan

    Lt Wilson took off and headed for Lakenheath AB

England. Steve waved to him from the hangar where the dusty

B-52 was parked. Steve would probably have one of the

structures guys put an ugly steel 90 degree patch on the

longeron and install it with jo-bolts. He wasn't even going

to write it up. He let the B-52 Program Office structure's

engineer know about it. Steve knew him because their desks

were about 50 feet apart back at Tinker. He also called

Robins AFB and let them know they needed a replacement

aileron shipped to Lakenheath for the F-15E.

    The commander told Steve he was rotating out and a new

ABDR engineer was coming in from Ogden with that team. Col

Smith shook Steve's hand and gave him his Commander's Round

Metal Object or RMO (A coin with the commanders name on

it). Steve didn't know the engineer from Ogden but he would

be well trained maybe not on the SASS but Steve was taking

that with him. He went to his tent to pack up to go home

the next day. Steve got to thinking these two deployments

puts my time away from home base at 178 days. *They are*

bringing me home before I can earn a short tour (a short

tour is 180 days or longer but shorter than 365 days, which

is a long tour). Those crappy bean counters at the

Pentagon. Why would they not want us to have the short

tour? I'll just break the plane tomorrow and fly out the

next day. Ah forget it. I'm getting out of here...tomorrow

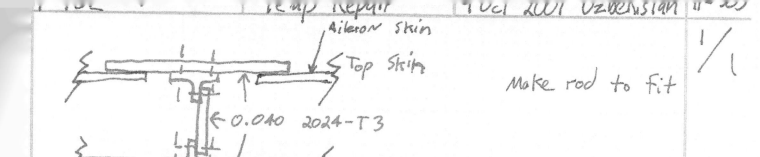

Aileron Skin

Top Skin

← 0.040 2024-T3

bottom Skin

Make rod to fit

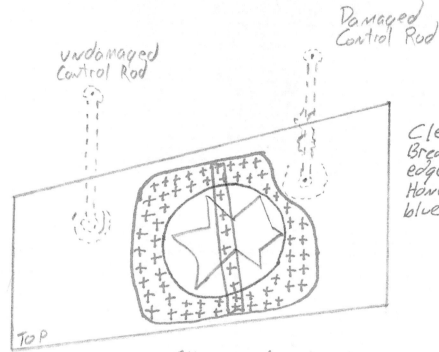

undamaged
Control Rod

Damaged
Control Rod

TOP

Clean Damage
Break all sharp
edges. Shave Al
Honeycomb inside
blue Circle

All repair sheets 2024-T3 0.040 thick. Make "I" beam from two "T" 7075-T6 extrusions with a 0.040 in thick Skin between them. Use hi-strength blind fasteners from group I MS90353 ¼ diameter with a 2D 4D 2D Fastener row pitch and spacing. Put fasteners in wet with sealant. Use same fasteners to install I beam to Skin

One time Ferry Flight to home base or depot for aileron replacement or repair. Minimal Fuel load No external ordinance (can have Air to Air Missiles)

49.

Monday October 14, 2001
Tinker AFB, Oklahoma, USA - The best Country in the World!

Steve walked down the ramp of the C-17. He walked over to the grass, knelt down and touched the ground. *It is good to be back in the USA! I'll have to go to Napoleon's sandwich shop across the street from the base or maybe Mr. Sprigs.* I could kill for some decent Barbeque.

Most people had a bunch of in-processing to do but not Steve because he was matrixed to the 654 CLSS. He walked out of Base ops and the commander of the 654 CLSS was there to take him over to the annex and drop off his mobility bags and weapon. He turned in his 9mm and bags and the

mobility NCOs said he could come back over in a week or two

to separate his personal gear from squadron gear. One of

the NCOs took him to his rental house and when he finally

got home he realized...he still had the SASS.

.

50.

Friday October 18, 2001
NW 23 and Penn, Oklahoma City

PARTY! It was good to get back home and see some friends. There was a party that was co-sponsored by Mac, one of Steve's church friends. It was a fun time. A lady that Steve knew, Michelle, had some girlfriends from Kingfisher OK that attended the party. Kingfisher was about 45 minutes from the metro and it was a country town. More like an Oakie town. When Steve saw this girl at the party he told Sergio, "I have to go talk to that girl." "Happy Hunting, I'm coming too." came the reply. Steve talked to the Kingfisher Girls and they were having a great

conversation…until Isaac the homeowner came over and was

clearly 3 sheets to the wind. Just plain drunk. Pretty much

everyone in the original conversation left the area and

Isaac was left jabbering to himself.

51.

Friday November 9, 2001
Quail Springs Mall

   The Crossings Community Church Singles Group organized
a movie night at the mall and Steve started scheming. How
could he see the Kingfisher girls again?

   At church on Sunday, Steve saw Michelle and he went up
to her and asked her to invite the Kingfisher girls to the
movie that Friday.

   When they went into the movies it didn't really work
out for Steve to sit next to Jenna so he felt a little
defeated. He decided to wait until they walked out of the

movies and see if he could get her phone number. Sergio walked out and Steve waved him over, "Hey bro I'm trying to get a girls phone number can you hang out so I don't look like a stalker?" Sergio was cool. He said, "Oh ya I can hang out. What girl are you trying to get digits for?" One of the Kingfisher girls, Jenna. "Oh good because I might ask out Sheena," replied Sergio. Sheena was Jenna's friend and was also a Kingfisher girl.

The girls walked out to Jenna's car and Steve told Sergio, "Catch you later Bro." Sergio said, "Good luck." Steve walked up to the driver's side and said, "Did you like the movie?" Both girls responded that they did. Steve looked at Jenna and said, "Can I have your phone number?" Jenna asked, "Whose phone number do you want?" Steve replied, "Yours Jenna. Do you have a pen?" Jenna said, I do not have a pen." Steve said, "OK I'll just have to remember it." She gave him the number and Steve replied, "Have a good night." "You too," they said. As Jenna was rolling the window up Steve thought he heard some giggles. He rushed back to his truck to find some sort of writing utensil because he didn't want to lose this number. He ended up finding a marker and wrote the number on his hand.

Tuesday November 13, 2001
Front porch of Steve's rental house

Steve was hanging out on his porch swing. *The wind must be out of the south* he thought because it was actually pleasant on his front porch. The wind…It always blows in Oklahoma. It is either a furnace blast of 100 degrees or a winter chill of 32 degrees. Steve thought he had let enough time lapse between asking for Jenna's number and actually calling her. So he dialed her number and she picked up. They talked a little about how their work weeks was going. She was a teacher at Kingfisher. Steve had told her a little about his week with some sergeant that had told him he was a sharp Lt and they didn't make many of those. She said, "You must be smart." Steve replied, "I am a genius…like Wiley Coyote." At first she thought he was being arrogant but then he mention Wiley Coyote. Wiley Coyote was a cartoon character that kept blowing himself up trying to eat the roadrunner in old cartoons. She laughed a little and Steve asked her out for the upcoming Friday. She accepted and they figured they would go out to a little movie theater in Kingfisher. They said their goodbyes and

Steve hung up. He heard that familiar voice of the Good Shepherd, "I will show you what a godly woman looks like that follows me." *Wow*, thought Steve. *This is not going to be a wild ride and that is what I need.*

Friday November 16, 2001
City Cafe Kingfisher OK

The Dodge Ram Steve owned was dirty. He spent the better part of the day washing it and cleaning out the inside. He met Jenna at the City Café in downtown Kingfisher. He brought her some red roses because hey, chicks dig flowers. They had a good dinner and it wasn't bad. Then they walked out to Steve's Truck and he opened the door for her. Jenna said, "Wow this is a nice clean truck." "Thanks, it wasn't," Steve replied. "It was overdue for some TLC."

They drove to the movies, parked and went in. When they entered the theater and sat down a student of Jenna's said hi to the couple. The student pulled Jenna over and whispered something to her. A few minutes later Jenna said her student said Steve was cute. They laughed a little because the student was ten. They had a fun date and Steve

dropped her off back at her car at the City Café.

Steve had about 45 minutes on the way back to his

house to contemplate the date. That was really fun. She

seems to be a stable godly woman and she is really pretty!

Is this what God was talking about?

Friday January 18, 2001
Rental house in the Village

Jenna and Steve had been dating for two months and

Steve kept praying that he would not mess up the

relationship. They had a date to a dinner and a movie and

went back to Steve's house to hang out. Thank fully Steve's

roommate was out with his girlfriend.

Jenna said she wanted to talk about something and

Steve knew deep down she was probably going to break up

with him. He was getting ready for a new assignment

somewhere away from Oklahoma and Jenna had always lived

there. She said, "I really have enjoyed dating you these

past couple of months. You have treated me with the utmost

respect. Most guys would have tried something by now. I

have been praying over our relationship and God has not

affirmed that we are supposed to move our relationship forward." What could Steve say? He hadn't heard the voice of God affirming their relationship either.

As far as breakups go this one was one of the better breakups. Steve was upset at losing Jenna but he knew God had great plans for him with providing a godly wife. Now when that would happen, Steve did not know.

52.

Wednesday March 14, 2002
Tinker AFB, OK

Lt Craken rolled into his cube a little late. He was getting short. His time here in the C/KC-135 Program office was drawing to a close. He was getting orders any day now. He put Schriever AFB CO as his number one pick of next assignment and his job would be in Space Command if he got that assignment. Number two was Ogden AFB UT in Material Command doing something similar to what he was doing now. He really wanted to be in the mountains where he could snowboard. The third choice was Patrick AFB in Space

Command in Florida.

He spun up his computer, opened email and let it churn

for three minutes so all the new emails would load. There

it was from Air Force Personnel Center… Next assignment… *OK*

here we go. I'm going to open it, thought Steve. *Wait*

before I open this I need to thank God the Father for so

much. Father God, I thank you for watching over me during

the deployment, battles and sniper engagements. Your will

be done not mine. Where ever I go for the Air Force I will

praise you and I will do your will, but for the record…I

love to snowboard!! Amen.

He opened the email…"COLORADO SPRINGS!!! Yes Yes!!!

First Space Operations Squadron launch Operations. Oh, that

is sooooo cool," Said Steve out loud to himself. His

coworkers were really stoked for him.

53.

This book was fun to write. Who knows if it will make any money. That was not the goal. The goal is to glorify God the Father, Jesus his son and the Holy Spirit and to entertain some peeps. Jesus has gotten me through some rough patches in life and He also can get you through anything you are going through. There is only way to heaven and that way is Jesus. John 14:6, "I am the way and the truth and the life. No one comes to the Father except through me…" If you are tired of the same old grind, of the stress of this life you can turn to Jesus. His yoke is easy and his burden is light. Let Jesus share your burdens.

If you would like eternal salvation it is easy to receive the free gift from Jesus. Romans 3:23 "For all have sinned and fallen short of the glory of God." I am a sinner and I am also now saved by grace by Jesus. The Bible also tells us what wages are, for sinning. They are death. Romans 5:8 says "…But God demonstrates His love for us in that while we were still sinners, Christ died for us."

Christianity is the only religion that I am aware of that our God died for us, to save us.

To receive this free gift of salvation all you have to do is follow Romans 10:9-10. Romans 10:9-10 "Confess with your mouth, 'Jesus is Lord', and believe in your heart that God raised him from the dead, you will be saved."

Below is a website you can go to learn more;

https://www.focusonthefamily.com/faith/what-must-i-do-to-be-saved/

If you liked the book drop me an email at;

Email: battledamge135@gmail.com

54.

Thanks to the following peeps: The 654 CLSS Squadron members from 1999-2002. Thank you to the C/KC-135 SPO members from 1999-2002. Thank you to Tom Ramsey for structures training and Karl. Thank you to Anne Tracy, Corey Cooper, Jason McGrogan, Jenny Moose, and Angie Butts for the comradery and banter. Thanks to my OKC church family; Dawnya, Adam, Ronda, Dionne, Mark, Jerry and Mike.

A special shout out to my parents. They are the best parents I could ever have. They gave me free rein in high school and supported me in college. They are great examples to model parenting after.

A Special shout out to best friend Kyle. He stuck by me through thick and thin, dates with girls and no dates with girls. We did get shot at by a crazy man and went through that together. He definitely made Starkville MS a fun place to be.

A double special thank you to my awesome wife that supports me in all I do no matter how crazy the scheme sounds.

Glossary

AFB Air Force Base - Where Aircraft and other Air Force offices are located.

AGE - Aerospace Ground Equipment. This provides power for the aircraft while the aircraft engines or Auxiliary Power unit (APU) are not running.

APU - a small turbine engine onboard most aircraft to provide power while on the ground instead of running the main engines and used to start the main engines

BBL - Beam Butt Line - Used to locate structure on aircraft in the "Z" direction or left and right as looking from the rear of the aircraft.

BS - Body Station - Used to locate structure on aircraft in the "X" direction.

CE - Civil Engineering - They take care of electricity, water, heating water and AC.

Chicom - Chinese Communists

CLSS - Combat Logistics Support Squadron.

DUL - Design Ultimate Loads - A method of structure repair that can be used when the engineer knows the design loads the manufacturer used in building the aircraft. This method is more efficient than MUL because the engineer can use less materials in the repair. It is also a less conservative repair than MUL.

EEOC Equal Employment Opportunity Counsel - They investigate discrimination complaints for the military.

EWO - Electronics Warfare Officer - The Officer in the backseat of a plane responsible for navigating and deploying countermeasures and offensive measures to impact the enemy.

FAA – Federal Aviation Administration. The Government organization that governs flight and aircraft certification.

FAM – Fighting Age Male

FOB – Forward Operating Base

F.O.D. – Foreign Object Damage – Can be anything ingested in an engine or other critical aircraft area. Such as a bolt left on the taxiway can cause significant internal damage to a jet engine.

GPS - Global Positioning System – the U. S. satellite based system used for precise timing and positioning

HVT – High Value Target

IED – Improvised Explosive Device

IFAK – (Individual First Aid Kit)
Karshi Khanabad – Secret U.S. base in Uzbekistan.

INS – Inertial Navigation System – a back-up positing system for aircraft where GPS is primary.

JFAC – Joint Forces Air Controller

OC-ALC - Oklahoma City Air Logistics Center - The OC-ALC is where Air Force aircraft come to depot maintenance every 4-6 years.

Lt. Lieutenant the first officer rank O1 (Second Lieutenant) and O2 (First Lieutenant).

MIA – Missing In Action

MSU - Mississippi State University

MUL – Material Ultimate Load – A method to repair structure using the material as a basis instead of design loads (see DUL)

NCO Non-Commissioned Offer. An enlisted member of the Air Force.

OPFOR – Opposing Forces

Ramp – The flat space that aircraft are parked outside next to taxi ways and runways.

R.O.T.C or ROTC is Reserve Officer's Training Corps. One of the three ways a person becomes an officer.

S.A.L.U.T.E. – A report about observed enemy actions. "S" Size, "A" Activity, "L" Location, "U" Uniform, "T" Time, "E" Equipment

T.O. – Technical Order. The manual for the aircraft that tells the technician how to repair the aircraft.

Snacko – The slang term used to denote an officer that is managing the base Morale Welfare and recreation (MWR). MWR covers the base bowling alley, theaters and gas stations.

SPO – System Program Office. The SPO manages weapon systems from the cradle to the grave.

TSgt – Technical Sergeant – An Air Force Rank, Enlisted E6 or sixth level of the enlisted rank.

UXO – Unexploded ordinance

WL – Water Line – The location tool to tell someone the height of something on an aircraft.

WS - Wing Station - The location tool to tell someone the location on the wing both right and left

Bibliography

Ebdon, Derek W. *Aircraft Battle Damage Repair Engineering Handbook*. 1. 2000th & 2010th ed. Vol. 1. Dayton, Oh: U.S. Air Force, 2000.

https://www.semanticscholar.org/paper/Reballasting-the-Kc-135-Fleet-for-Fuel-Efficiency-Morrison/bef40192795a51091ac3462de0b39511196baa68   Figure 1.

https://www.semanticscholar.org/paper/Reballasting-the-Kc-135-Fleet-for-Fuel-Efficiency-Morrison/bef40192795a51091ac3462de0b39511196baa68   Figure 2.

B52 load out https://airwingmedia.com/downloads/b52-stratofortress/

B52 skeleton

https://www.pinterest.com/pin/524528687824562795/

https://www.pinterest.com/pin/466544842622965428/visual-search/?imageSignature=4b05e5cdd8849522900b709431d0def0

AN/ALE-50 towed decoy system
https://en.wikipedia.org/wiki/AN/ALE-50_towed_decoy_system#:~:text=The%20AN%2FALE%2D50%20towed%20decoy%20system%20is%20an%20anti,non%2DUnited%20States%20air%20forces.

Air launched decoy
https://en.wikipedia.org/wiki/ADM-160_MALD

Made in the USA
Coppell, TX
13 February 2025

45886262R00247